TIME-DELAY SYSTEMS
Stability and Performance Criteria with Applications

MATHEMATICS AND ITS APPLICATIONS
Series Editor: G. M. BELL
Emeritus Professor of Mathematics, King's College London, University of London

STATISTICS, OPERATIONAL RESEARCH AND COMPUTATIONAL MATHEMATICS Section
Editor: B. W. CONOLLY,
Emeritus Professor of Mathematics (Operational Research), Queen Mary College, University of London

Mathematics and its applications are now awe-inspiring in their scope, variety and depth. Not only is there rapid growth in pure mathematics and its applications to the traditional fields of the physical sciences, engineering and statistics, but new fields of application are emerging in biology, ecology and social organization. The user of mathematics must assimilate subtle new techniques and also learn to handle the great power of the computer efficiently and economically.

The need for clear, concise and authoritative texts is thus greater than ever and our series endeavours to supply this need. It aims to be comprehensive and yet flexible. Works surveying recent research will introduce new areas and up-to-date mathematical methods. Undergraduate texts on established topics will stimulate student interest by including applications relevant at the present day. The series will also include selected volumes of lecture notes which will enable certain important topics to be presented earlier than would otherwise be possible.

In all these ways it is hoped to render a valuable service to those who learn, teach, develop and use mathematics.

Mathematics and its Applications
Series Editor: G. M. BELL
Professor of Mathematics, King's College London, University of London

Anderson, I.	COMBINATORIAL DESIGNS: Construction Methods
Artmann, B.	CONCEPT OF NUMBER: From Quaternions to Monads and Topological Fields
Arczewski, K. & Pietrucha, J.	MATHEMATICAL MODELLING OF COMPLEX MECHANICAL SYSTEMS: Volume 2: Continuous Models
Arczewski, K. & Pietrucha, J.	MATHEMATICAL MODELLING OF COMPLEX MECHANICAL SYSTEMS: Volume 1: Discrete Models
Bainov, D.D. & K. Covachev	THE AVERAGING METHOD AND ITS APPLICATIONS
Bainov, D.D. & Simeonov, P.S.	SYSTEMS WITH IMPULSE EFFECT: Stability, Theory and Applications
Baker, A.C. & Porteous, H.L.	LINEAR ALGEBRA AND DIFFERENTIAL EQUATIONS
Balcerzyk, S. & Jöseflak, T.	COMMUTATIVE RINGS
Balcerzyk, S. & Jöseflak, T.	COMMUTATIVE NOETHERIAN AND KRULL RINGS
Baldock, G.R. & Bridgeman, T.	MATHEMATICAL THEORY OF WAVE MOTION
Ball, M.A.	MATHEMATICS IN THE SOCIAL AND LIFE SCIENCES: Theories, Models and Methods
Barnett, S.	SOME MODERN APPLICATIONS OF MATHEMATICS
Bartak, J., Herrmann, L., Lovicar, V. & Vejvoda, D.	PARTIAL DIFFERENTIAL EQUATIONS OF EVOLUTION
Bejancu, A.	FINSLER GEOMETRY AND APPLICATIONS
Bell, G.M. & Lavis, D.A.	STATISTICAL MECHANICS OF LATTICE MODELS, Vols. 1 & 2
Berry, J.S., Burghes, D.N., Huntley, I.D., James, D.J.G. & Moscardini, A.O.	MATHEMATICAL MODELLING COURSES
Berry, J.S., Burghes, D.N., Huntley, I.D., James, D.J.G. & Moscardini, A.O.	MATHEMATICAL MODELLING METHODOLOGY, MODELS AND MICROS
Berry, J.S., Burghes, D.N., Huntley, I.D., James, D.J.G. & Moscardini, A.O.	TEACHING AND APPLYING MATHEMATICAL MODELLING
Brown, R.	TOPOLOGY: A Geometric Account of General Topology, Homotopy Types and the Fundamental Groupoid
Burghes, D.N. & Borrie, M.	MODELLING WITH DIFFERENTIAL EQUATIONS
Burghes, D.N. & Downs, A.M.	MODERN INTRODUCTION TO CLASSICAL MECHANICS AND CONTROL
Burghes, D.N. & Graham, A.	INTRODUCTION TO CONTROL THEORY, INCLUDING OPTIMAL CONTROL
Burghes, D.N. & Wood, A.D.	MATHEMATICAL MODELS IN THE SOCIAL, MANAGEMENT AND LIFE SCIENCES
Butkovskiy, A.G.	GREEN'S FUNCTIONS AND TRANSFER FUNCTIONS HANDBOOK
Cartwright, M.	FOURIER METHODS: for Mathematicians, Scientists and Engineers
Cerny, I.	COMPLEX DOMAIN ANALYSIS
Chorlton, F.	VECTOR AND TENSOR METHODS
Cohen, D.E.	COMPUTABILITY AND LOGIC
Cordier, J.-M. & Porter, T.	SHAPE THEORY: Categorical Methods of Approximation
Crapper, G.D.	INTRODUCTION TO WATER WAVES
Cross, M. & Moscardini, A.O.	LEARNING THE ART OF MATHEMATICAL MODELLING
Cullen, M.R.	LINEAR MODELS IN BIOLOGY
Dunning-Davies, J.	MATHEMATICAL METHODS FOR MATHEMATICIANS, PHYSICAL SCIENTISTS AND ENGINEERS
Eason, G., Coles, C.W. & Gettinby, G.	MATHEMATICS AND STATISTICS FOR THE BIOSCIENCES
El Jai, A. & Pritchard, A.J.	SENSORS AND CONTROLS IN THE ANALYSIS OF DISTRIBUTED SYSTEMS
Exton, H.	MULTIPLE HYPERGEOMETRIC FUNCTIONS AND APPLICATIONS
Exton, H.	HANDBOOK OF HYPERGEOMETRIC INTEGRALS
Exton, H.	q-HYPERGEOMETRIC FUNCTIONS AND APPLICATIONS

Series continued at back of book

TIME-DELAY SYSTEMS
Stability and Performance Criteria with Applications

J.E. MARSHALL
University of Bath, UK
H. GÓRECKI
IAIST, AMM, Krakow, Poland
A. KORYTOWSKI
IAIST, AMM Krakow, Poland
K. WALTON
The University of Bath, UK

ELLIS HORWOOD
NEW YORK   LONDON   TORONTO   SYDNEY   TOKYO   SINGAPORE

First published in 1992 by
ELLIS HORWOOD LIMITED
Market Cross House, Cooper Street,
Chichester, West Sussex, PO19 1EB, England

A division of
Simon & Schuster International Group
A Paramount Communications Company

© Ellis Horwood Limited, 1992

All rights reserved. No part of this publication may be reproduced, stored in a retrieval system, or transmitted, in any form, or by any means, electronic, mechanical, photocopying, recording or otherwise, without the prior permission, in writing, of the publisher

Printed and bound in Great Britain
by Redwood Press, Melksham

British Library Cataloguing in Publication Data

A catalogue record for this book is available from the the British Library

0–13–465923–6

Library of Congress Cataloging-in-Publication Data

Available from the publisher

Table of contents

Preface ix

1 Performance criteria for delay-free systems 1

  1.1 Introduction 1
  1.2 Integral performance criteria 2
  1.3 Delay-free systems 3
  1.4 Evaluation of performance criteria: time-domain methods 6
  1.5 Evaluation of performance criteria: complex variable methods 8
  1.6 Evaluation of performance criteria: matrix methods 11
  1.7 Review of the later chapters 14

2 Time-delay systems and stability 16

  2.1 Introduction 16
  2.2 Systems with a single time delay 17
  2.3 Method of steps in the transform domain 18
  2.4 Stability of time-delay systems with a single delay 19
  2.5 Stability windows 29
  2.6 Systems with more than one delay: commensurate delays 30
  2.7 Stability test for commensurate delays 31
  2.8 Commensurate delays and the method of steps 37

3 The method using a generalized Heaviside expansion 38

  3.1 Introduction 38
  3.2 The single-delay problem: a simple example 39
  3.3 The single-delay problem: the general case 42
  3.4 Extension to other integral criteria 47

4 The method based on Parseval's theorem and contour integration — 51

- 4.1 Introduction — 51
- 4.2 The single-delay problem: a simple example — 52
- 4.3 The single-delay problem: the general case — 55
- 4.4 Extension to two commensurate delays, h and $2h$ — 59
- 4.5 Several commensurate delays — 65
- 4.6 Extension to weighted quadratic criteria — 71
- 4.7 Stored functions, arbitrary inputs and initial conditions — 75

5 The Lyapunov method — 83

- 5.1 Introduction — 83
- 5.2 Case of a single delay and pointwise initial conditions — 89
- 5.3 Solution of the Lyapunov system of equations — 91
- 5.4 Example — 97
- 5.5 Case of arbitrary initial conditions — 101
- 5.6 Performance criteria with weighting functions and derivatives — 104
- 5.7 Systems with many delays — 110
- 5.8 Neutral systems — 112

6 Evaluation of integrals for sampled-data systems — 121

- 6.1 Introduction — 121
- 6.2 Review of z-transform results — 121
- 6.3 Cost functionals for sampled-data systems — 125
- 6.4 Stability of a sampled-data system — 126
- 6.5 Stability for general N — 127
- 6.6 Relationships between the discrete and the continuous Parseval integrals — 129
- 6.7 Extension of Sklansky's method for large N — 131
- 6.8 Evaluation of integrals of the mixed type arising in sampled-data systems — 135

7 All-pass systems — 143

- 7.1 Introduction — 143
- 7.2 Stability of all-pass systems — 145
- 7.3 Cost functional evaluation of an all-pass system — 148
- 7.4 An all-pass cost-difference theorem — 152
- 7.5 Exponential inputs to all-pass systems — 155
- 7.6 Padé approximants of the exponential function — 156
- 7.7 Padé approximant and stability — 157
- 7.8 Padé approximant and the evaluation of cost functionals — 160

8 Application to PID control — 163

8.1 Time-delay systems with PID-controllers — 163
8.2 Optimization of conventional control systems with first-order plant and unit step input — 172
8.3 Conventional control system activated by initial conditions — 180
8.4 Application to stochastic inputs — 190
8.5 Conventional control system driven by white Gaussian noise — 196

9 Predictive control: mismatch problems — 201

9.1 Introduction — 201
9.2 A predictor control scheme — 202
9.3 Mismatch — 204
9.4 Mismatch and performance — 205
9.5 Sensitivity and the sign of mismatch: the temporal case — 207
9.6 The parametric sensitivity coefficient — 210
9.7 Cross-sensitivity — 211
9.8 Higher derivatives — 212
9.9 Summary — 213

10 Predictive control: exact cost functionals — 214

10.1 Introduction — 214
10.2 Parametric mismatch: stability — 215
10.3 Parametric mismatch: cost — 217
10.4 Temporal mismatch: stability — 218
10.5 Temporal mismatch: $h_0 = 0$ — 221
10.6 Evaluation of J when $D(s) = 0$ and $h_0 = 0$ — 225
10.7 Commensurate delays — 227
10.8 Padé all-pass model in a predictor scheme — 231
10.9 Summary — 236

References — 238

Index — 242

Preface

The analysis and control of linear delay-free dynamical systems has a wide literature, is well understood and forms part of the curriculum of most engineers and many applied mathematicians.

In this book, the analysis of stability and performance of linear systems with time delay is carried forward using as far as possible the same mathematical tools, i.e. Laplace transform, Lyapunov methods and frequency domain complex variable methods. Our intention is to provide exact analytical solutions as far as this is possible in the analysis of stability and performance qualified in terms of cost functionals, typically, but not exclusively, the integral of squared error.

One reason for the relative simplicity in the analysis of delay-free systems is the finite number of roots of the characteristic equation, i.e. the finite number of system poles. This simplifying feature is lost when time delays are introduced. This book derives and describes methods which address this new problem: that the number of system poles is no longer finite.

The methods to be described have their origin in a decade of cooperation between two research groups, in Krakow at the Institute of Automatics in the Academy of Mining and Metallurgy, and in the School of Mathematical Sciences at the University of Bath. These methods show how the infinite-dimensional problems of time-delay systems may be reduced, in many cases, to investigating the properties of finite polynomials. There is a happy convergence of two methods for cost functional evaluation, one Lyapunov-theory based and the other complex-variable based, to the consideration of the same finite polynomial. Certain of the roots of this same polynomial are essential in the corresponding stability analysis. Other features of the book are an exploration of some related problems in discrete systems, and all-pass systems, and the book concludes with three chapters of applications, with fully worked examples.

We are happy to record our thanks to the Royal Society, the Polish Academy of Sciences and the British Council, who have encouraged and supported our collaboration.

Research collaboration is enriched by discussions with one's colleagues and in this regard we wish to thank the following for their contribution:

Alan Barnes, Pat Davis, Stanislaw Fuksa, Piotr Grabowski, Peter Hydon, Bryan Ireland.

We thank Sue Collins, Liz Gaynor, Hazel Gott, Sally Waldegrave and Yasuko Thorn for their accurate and very patient word-processing skills.

Henryk Górecki	John Marshall
Adam Korytowski	Keith Walton
AMM Kraków	University of Bath

1

Performance criteria and delay-free systems

1.1 INTRODUCTION

The control of linear delay-free continuous-time systems has a rich research and tutorial literature and is presented widely to undergraduate audiences. The extension of this classical theory even to linear systems with delay is absent from many standard control texts. This omission, though surprising in that many processes are dominated by delay, is nevertheless understandable in that it is held that the extension of delay-free to delay cases leads to difficult analytical problems, some of which are to be discussed in Chapter 2.

Until recently, no exact solutions to such problems were available and the only alternative was to use approximate solutions. However, this is no longer the case, and this book is concerned with showing that many delay-free methods can be extended to the control of systems with time delay.

It is important to stress that we are seeking exact solutions to time-delay problems, where this is possible, not approximate ones.

The theme of the book is that where exact solutions are possible they should be sought. By exact solutions we imply solutions not involving infinite summations, or numerical solutions. When analytical solutions have been found (and this is not always possible even in principle) it will be possible to explore and explain the basis of techniques hitherto found by experience or well-informed intuition.

In its rapid growth since the 1940s control theory has continued to develop along two major routes. The Laplace transform is an obvious and productive tool for the analysis of linear time-invariant systems. The other is the use of matrix-based methods, where tools of linear algebra and state-space methods have been used powerfully in combination. It is now known that the Laplace transform complex variable techniques and the 'modern' methods combine happily to provide very powerful tools for the analysis and control of linear systems.

The book's main thrust is towards continuous control of continuous processes but some discrete techniques are nevertheless found useful and are to be discussed later.

1.2 INTEGRAL PERFORMANCE CRITERIA

Comparisons between rival control strategies are made with reference to an agreed criterion of performance, or cost. A practical criterion for low-order systems is that of overshoot to a step response, beyond a steady-state constant value. Rise-time, i.e. the time for a step response to achieve some nominal percentage of its final value, is also a practical criterion. These point-wise criteria are recognised as restrictive. It is better to use several such criteria in conjunction or to use criteria which assess quality over a broad time-span.

The most widely used criterion or cost functional in linear systems theory, with parallels in many sciences, is that of the integral of a squared error (ISE). It is helpful to think of error as the difference between an actual and a desired response, but other definitions are possible and useful. The ISE criterion has found wide use for practical and theoretical reasons. We shall review methods for the evaluation of ISE for delay-free systems in later sections of this chapter, choosing largely those methods which find extension to the time-delay case in later chapters. We do not restrict ourselves solely to ISE, but choose to stress this criterion initially for comparison of methods, before demonstrating extensions to other criteria.

The ISE criterion is extended in various ways. The most obvious is a weighted sum of such integrals, where the weights themselves may be regarded, *ab initio*, as design variables.

Other cost functionals with polynomially or exponentially weighted versions of error squared or error derivatives are possible.

Criteria in which are used the absolute value (modulus) of error, or other variables, are less amenable to non-numerical treatment. An excellent review paper by Graham and Lathrop (1953), comparing the dependence of various criteria on a single design parameter, is very instructive. The paper demonstrates how a parameter which minimizes the value of one criterion often comes close to minimizing others.

Excellent reviews of performance criteria, their historical development, and motivation are to be found in the paper of Fuller (1967) and in the book by Jury (1974, Chapters. 4 and 5).

An important class of 'classical' control techniques known as 'analytical design' concerns the minimization of performance criteria with respect to 'controller' parameters. In analytical design it is assumed that the controller structure is chosen in advance, but that the values of certain coefficients are freely disposable for control purposes. If a simple analytical expression is available in terms of system and control parameters then minimization of the chosen criterion (not necessarily ISE) can proceed by way of standard methods of calculus and algebra. We note that the choice of controller structure and the choice of criterion depend on the designer's judgement, or on a given specification.

Analytical expressions, particularly for ISE, have been available for linear delay-free systems since the 1940s, in tabular form. The books of James, Nichols and Phillips (1947), and Newton, Gould and Kaiser (1957) are the classical references to analytical design. More recently, Åström (1970) and Jury (1974) have provided new methods.

There is a strong connection between criteria of performance and stability. If a system is unstable in the sense that an associated error function fails to tend asymptotically to zero after the removal of a stimulus, it is clear that the criterion of performance will not take finite values. Further, the integral of a squared error, or any other positive function, must undoubtedly produce a cost of positive sign. The positivity of such criteria is a property that can be exploited in stability studies.

In the light of this comment it is not surprising that the Lyapunov stability methods, which date from his seminal non-linear stability paper of 1895, should prove of use in the evaluation of quadratic, and other criteria. Linear systems have associated quadratic Lyapunov functions. This connection is particularly helpful when systems are analysed using matrix methods (state-space methods). We shall give examples of Lyapunov methods following the exposition of MacFarlane (1963), quoted by Jury (1974). An excellent survey of Lyapunov techniques, of general application, is given by Kalman and Bertram (1960). The connections between stability, which is discussed in detail in Chapter 2, and the evaluation of performance integrals in Chapters 3, 4, 5 and 6, are an important theme of the book.

1.3 DELAY-FREE SYSTEMS

We shall review the notation standard to Laplace transform techniques in the control context, and matrix techniques for the analysis of constant parameter linear systems. Of particular importance are the Heaviside expansion formulae, error transfer function, stability, and in the matrix formulation the fundamental matrix and its properties.

Excellent introductions to such techniques are to be found in the books of Truxal (1958), Barnett (1975), Jacobs (1974), MacFarlane (1970), Rosenbrock and Storey (1970). Ogata (1987), Barnett (1984) and Rosenbrock (1970, 1974) are particularly to be recommended for the state-space methods.

For time-invariant linear systems, the response of an initially undisturbed system to a Dirac delta function, $\delta(t)$, is well defined and is called the impulse response, denoted by $g(t)$. Output, $y(t)$, and input, $x(t)$, for these systems are related by the convolution integral

$$y(t) = \int_{-\infty}^{+\infty} x(\lambda)g(t - \lambda)\, d\lambda \tag{1.1}$$

For causal systems, $g(t) = 0$ for $t < 0$ so for inputs which are zero for negative t, this may be written

$$y(t) = \int_{0}^{t} x(\lambda)g(t - \lambda)\, d\lambda \tag{1.2}$$

The convolution theorem of the Laplace transform enables us to write

$$Y(s) = G(s)X(s), \tag{1.3}$$

where $Y(s)$, $G(s)$, $X(s)$ are the Laplace transforms of $y(t)$, $g(t)$, $x(t)$ respectively. $G(s) = Y(s)/X(s)$ is the transfer function of the system.

For delay-free systems described by an nth-order ordinary differential equation with constant coefficients the transfer function takes the form $G(s) = P(s)/Q(s)$ where $P(s)$ and $Q(s)$ are polynomials in s of order m and n respectively and for the system to be physically realizable, $m \leqslant n$. In their factored form

$$P(s) = K \prod_{i=1}^{m} (s - z_i), \quad Q(s) = \prod_{j=1}^{n} (s - p_j), \tag{1.4}$$

where z_i, the zeros of $G(s)$, are the roots of $P(s)$ and p_j, the poles of $G(s)$ are the roots of $Q(s)$ and the constant K is usually taken to be positive. As the coefficients in the polynomials $P(s)$, $Q(s)$ are real, zeros and poles when complex occur in conjugate pairs, which as with the real poles and zeros may be repeated.

Writing $G(s)$ in its partial fraction form, and assuming for now no repeated factors we may write

$$G(s) = \sum_{j=1}^{n} \frac{A_j}{s - p_j}$$

so that

$$g(t) = \mathscr{L}^{-1} G(s) = \sum_{j=1}^{n} A_j \exp(p_j t) H(t). \tag{1.5}$$

This is the Heaviside expansion formula for $g(t)$. In the case of $y(t)$ similarly expressed there will be terms from the poles of the input, and the system poles.

When unity negative feedback is applied to the system with open-loop transfer function $G(s) = P(s)/Q(s)$ with error defined as input minus output, the closed-loop transfer function is given by $P(s)/(P(s) + Q(s))$, and the transform error by

$$E(s) = \frac{Q(s)X(s)}{P(s) + Q(s)}. \tag{1.6}$$

$P(s) + Q(s) = 0$ is the closed-loop characteristic equation, and its n roots are the closed-loop system poles. $E(s)$, the Laplace transform of the error, will figure largely in later chapters.

If the nth-order ordinary differential equation is written as a system of first-order differential equations (or as in cases where there are several inputs and outputs) then the matrix description is the more natural. This will take the form

$$\frac{dx}{dt} = Ax + Bu, \quad y = Cx, \tag{1.7}$$

where u and y are input and output vectors, and x is the state vector. The dimensions of u and y are the numbers of inputs and outputs respectively. A, B, C are matrices of appropriate dimensions. For the single-input single-output nth-order system, B and C are column and row n vectors respectively, and A is clearly square, of dimension n.

In the case of linear state feedback, with the control law $u = Fy$, (1.7) gives

$$\frac{dx}{dt} = Ax + BFy = [A + BFC]x$$

$$= A'x,$$ say, which we note to be a special case of the same equation, with A replaced by A', and $u(t)$ zero.

The standard form of the solution to the system of equations (1.7) in general is

$$x(t) = \exp(At)x(0) + \int_0^t \exp(A(t - t'))Bu(t')\, dt' \qquad (1.8)$$

$$y(t) = Cx(t)$$

where $x(0)$ is the initial value of $x(t)$.

This solution may be derived by Laplace transform, or eigenvector techniques.

There are two special cases to consider:

(a) when $x(0) = 0$

$$x(t) = \int_0^t \exp(A(t - t'))Bu(t')\, dt'. \qquad (1.9)$$

Hence the impulse response for the single-input single-output case when the scalar $u(t)$ is replaced by $\delta(t)$ is obtained from

$$x(t) = \int_0^t \exp(A(t - t'))B\delta(t')\, dt' = \exp(At)B, \qquad (1.10)$$

and the impulse response is given by $g(t) = C \exp(At)B$.

(b) When $u(t) = 0$, $x(t) = \exp(At)x(0)$. \qquad (1.11)

It is usual to replace $\exp(At)$ by $\Phi(t)$, called the fundamental matrix, where clearly $\Phi(t)$ satisfies

$$\frac{d\Phi(t)}{dt} = A\Phi(t); \quad \Phi(0) = I_n,$$

where I_n is the unit matrix of order n. It may be shown that $A\Phi = \Phi A$.

In terms of the eigenvalues λ_i, and the eigenvectors x_i of A, we may write (1.11) as

$$x(t) = \sum_{i=1}^{n} \alpha_i x_i \exp(\lambda_i t), \qquad (1.12)$$

where α_i is the ith element of the column vector $Q^{-1}x(0)$, where Q denotes the matrix with columns the eigenvectors x_i. We have assumed non-repeated eigenvalues, for convenience.

Stability properties are immediate from the Heaviside expansion (1.5), in terms of poles, or (1.12) in terms of eigenvalues. In each case an asymptotically stable system will result when the exponents in these expressions, i.e. the system poles, or equivalently the system eigenvalues, have negative real parts.

However, it is helpful to have stability methods that do not require the explicit solution of the eigenvalue problem, or explicit values for system poles.

There are many such stability methods; we mention two which we shall make use of later: the Routh–Hurwitz criteria, which we shall describe and give a short proof of later in this chapter and a Lyapunov technique. The Lyapunov result is that the system $\dot{x} = Ax$ is asymptotically stable provided that there exists a positive definite matrix P as the solution to the continuous Lyapunov equation

$$A^T P + PA = -V \tag{1.13}$$

for any given real symmetric positive definite matrix V.

An excellent discussion of stability theory from both points of view is given by Barnett (1975, Chapter 5).

1.4 EVALUATION OF PERFORMANCE CRITERIA: TIME-DOMAIN METHODS

Methods for the evaluation of

$$J = \int_0^\infty e^2(t) \, dt, \tag{1.14}$$

and generalizations of this integral for the delay-free case, are to be developed in this and the next two sections. Later in this section we shall show that it is possible to evaluate J, without explicit solution of the differential equation for $e(t)$.

However, we begin by assuming that the solution for $e(t)$ is known and expressed in the form of the expansion equation (1.5). For simplicity we take the non-repeated pole case, i.e.

$$e(t) = \sum_{i=1}^n A_i \exp(p_i t), \tag{1.5}$$

with all p_i in the open left half-plane. In the case $n = 1$, $p_1 < 0$ we have simply

$$J = \int_0^\infty A_1^2 \exp(2p_i t) \, dt = -\frac{A_1^2}{2p_1} > 0. \tag{1.16}$$

A helpful alternative is to use the corresponding differential equation directly: i.e.

$$\frac{de(t)}{dt} = -ae(t),$$

say, where $a = -p_1$, and $e(0) = A_1$. Then

$$\int_0^\infty e^2(t)\,dt = -\int_0^\infty \frac{1}{a}\frac{de(t)}{dt} e(t)\,dt = -\int_0^\infty \frac{1}{a}\frac{d}{dt}\left(\frac{e^2(t)}{2}\right) dt$$

$$= \frac{e^2(0)}{2a}, \qquad (1.17)$$

in agreement with the above.

This method has a powerful extension.

Returning to the case of $n \neq 1$, we note that

$$J = \int_0^\infty e^2(t)\,dt = \int_0^\infty \sum_{i=1}^n A_i e^{p_i t} \sum_{j=1}^n A_j e^{p_j t}\,dt$$

$$= \int_0^\infty \sum_{i=1}^n \sum_{j=1}^n A_i A_j e^{(p_i+p_j)t}\,dt = -\sum_{i=1}^n \sum_{j=1}^n \frac{A_i A_j}{(p_i+p_j)} \qquad (1.18)$$

If the p_I are explicitly known then these summations may be evaluated. On the other hand, if they are not known explicitly it is known that they are the zeros of the characteristic equation. Using this fact, James, Nichols and Phillips were able to obtain simple expressions for such summations involving the coefficients of this equation. In this manner it is possible to evaluate the J.

A technique due to Allwright (1980) enables other integrals to be found, by an extension of the first-order method just shown. We take a second-order example,

$$a_0 \frac{d^2 e(t)}{dt^2} + a_1 \frac{de(t)}{dt} + a_2 e(t) = 0,$$

where $e(0) = e_0$, and $\dot{e}(0) = e_1$.

The extension is obvious, and as shown in Allwright's paper leads to another 'short' proof of the Routh–Hurwitz method.

Define

$$J_0 = \int_0^\infty e^2(t)\,dt, \quad J_1 = \int_0^\infty \left(\frac{de(t)}{dt}\right)^2 dt, \qquad (1.19)$$

and note the following results:

$$\int_0^\infty e(t)\frac{de(t)}{dt}\,dt = \frac{1}{2}\int_0^\infty \frac{d}{dt}(e^2(t))\,dt = -\frac{e_0^2}{2} \qquad (1.20)$$

$$\int_0^\infty \frac{de(t)}{dt} \frac{d^2e(t)}{dt^2} dt = \frac{1}{2}\int_0^\infty \frac{d}{dt}\left(\frac{de(t)}{dt}\right)^2 dt = -\frac{e_1^2}{2} \qquad (1.21)$$

$$\int_0^\infty e(t) \frac{d^2e(t)}{dt^2} dt = \left[e(t)\frac{de(t)}{dt}\right]_0^\infty - \int_0^\infty \left(\frac{de(t)}{dt}\right)^2 dt \qquad (1.22)$$

$$= -e_0e_1 - J_1. \qquad (1.23)$$

Multiply the differential equation by $e(t)$, and integrate over the usual range to obtain, via the results above,

$$a_0(-e_0e_1 - J_1) - a_1\frac{e_0^2}{2} + a_2 J_0 = 0. \qquad (1.24)$$

Similarly, multiplying by $\dot{e}(t)$ and integrating we obtain

$$-a_0\frac{e_1^2}{2} + a_1 J_1 - a_2\frac{e_0^2}{2} = 0. \qquad (1.25)$$

These equations are simultaneous equations for J_0, J_1 in terms of the coefficients a_0, a_1, a_2, and the initial values e_0, e_1, which may be solved to give $J_0 = \{(a_0a_2 + a_1^2)e_0^2 + 2a_0a_1e_0e_1 + a_0^2e_1^2\}/2a_1a_2$ and $J_1 = (a_0e_1^2 + a_2e_0^2)/2a_1$.

This result extends to the more general case, where $e(t)$ satisfies an nth-order equation, with $e(t)$ and its first $n-1$ derivatives known at $t = 0$. An extension of the method may also be used for integrands such as $t^n e^2(t)$ etc.

1.5 EVALUATION OF PERFORMANCE CRITERIA: COMPLEX VARIABLE METHODS

These methods are applicable when $E(s)$, the Laplace transform of error, is known. We seek a method that does not require an explicit time-domain solution, and will extend to the time-delay case. We shall use Parseval's theorem for the Laplace transform, some partial fraction results that exploit the symmetry properties of the integrals of this theorem, and a standard result, which we shall prove, from contour integration.

Parseval's theorem is stated for our purposes in the following way:

Given that the Laplace transform of $e(t)$ and

$$\int_0^\infty e^2(t)\, dt$$

exist then

$$\int_0^\infty e^2(t)\, dt = \frac{1}{2\pi i}\int_{-i\infty}^{+i\infty} E(s)E(-s)\, ds. \qquad (1.26)$$

Sec. 1.5] Evaluation of performance criteria: complex variable methods 9

The evaluation that we shall use follows that of Newton, Gould and Kaiser (1957), with some minor changes of notation.

The standard complex variable result that we require is now introduced. Suppose, for example, we wish to evaluate

$$I = \frac{1}{2\pi i} \int_{-i\infty}^{i\infty} \frac{F(s)}{A(s)} \, ds, \qquad (1.27)$$

where $A(s) = a_0 s^n + a_1 s^{n-1} + \ldots + a_n$ and $F(s) = f_1 s^{n-1} + f_2 s^{n-2} + \ldots + f_n$, where we are given that all the roots of $A(s)$ lie in the negative half-plane, and $a_0 \neq 0$. Consider Γ, the closed contour, (D-contour), formed by the imaginary axis and the infinite semi-circle enclosing the positive half-plane. Owing to the properties of $A(s)$ there are no poles within the contour, so that the integral

$$\frac{1}{2\pi i} \int_{\Gamma} \frac{F(s)}{A(s)} \, ds = 0. \qquad (1.28)$$

Hence

$$\frac{1}{2\pi i} \int_{-i\infty}^{i\infty} \frac{F(s)}{A(s)} \, ds + \frac{1}{2\pi i} \int_{C_R} \frac{F(s)}{A(s)} \, ds = 0 \qquad (1.29)$$

where the second integral follows the RH infinite semi-circle in a clockwise direction. On the infinite semi-circle

$$F(s)/A(s) \to \frac{f_1}{a_0 s} \qquad (1.30)$$

so that the second integral is

$$-\frac{f_1}{2a_0},$$

the negative sign arising from the clockwise rotation. Hence

$$\frac{1}{2\pi i} \int_{-i\infty}^{i\infty} \frac{F(s)}{A(s)} \, ds = \frac{f_1}{2a_0}. \qquad (1.31)$$

Hence given integrands of the form of this example, and their obvious generalization to higher order, we see that the integral may be found by inspection of $F(s)$ and $A(s)$ provided that all zeros of $A(s)$ have negative real parts.

We now use this result in Parseval's theorem. We express the Laplace transform of error, for example that of a unity negative feedback system, as

$$E(s) = \frac{B(s)}{A(s)} = \frac{b_1 s^{n-1} + b_2 s^{n-2} + \ldots + b_n}{a_0 s^n + a_1 s^{n-1} + \ldots + a_n}, \quad a_0 \neq 0 \qquad (1.32)$$

where the polynomial $B(s)$ is of lower order than the polynomial $A(s)$, as required for J to exist, as may be seen from Parseval's theorem.

Exploiting symmetry we express the integrand $E(s)E(-s)$ in the form:

$$E(s)E(-s) = \frac{B(s)}{A(s)} \frac{B(-s)}{A(-s)} = \frac{F(-s)}{A(s)} + \frac{F(s)}{A(-s)} \tag{1.33}$$

where $F(s)$ is a polynomial of degree less than $A(s)$.

$F(s)$ may be written for convenience

$$F(s) = (-)^{n-1}\{f_1 s^{n-1} + f_2 s^{n-2} + \ldots + f_{n-1} s + f_n\},$$

and

$$A(s)F(s) + A(-s)F(-s) = B(s)B(-s). \tag{1.34}$$

Equating coefficients to find the coefficients of $F(s)$, the following set of equations is found:

$$\begin{bmatrix} a_1 & a_0 & 0 & 0 & & \cdot & & \cdot \\ a_3 & a_2 & a_1 & a_0 & 0 & & \cdot & \\ a_5 & a_4 & a_3 & & & & & \\ \cdot & & & & & & & \\ \cdot & & & & & & & \\ \cdot & & & & & & & \\ & & & & a_n & a_{n-1} & a_{n-2} & \\ & & & & & & a_n & \end{bmatrix} \begin{bmatrix} f_1 \\ f_2 \\ \cdot \\ \cdot \\ \cdot \\ f_n \end{bmatrix} = \begin{bmatrix} g_1 \\ g_2 \\ \cdot \\ \cdot \\ \cdot \\ g_n \end{bmatrix} \tag{1.35}$$

where

$$g_r = \frac{1}{2} \sum_{m=1}^{2r-1} (-)^{m-1} b_{2r-m} b_m \tag{1.36}$$

and $b_m = 0$ for $m > n$.

In principle, all the f coefficients may be found via Cramer's rule, but by the earlier result we require only f_1.

Note that this array is the same as that of the Routh–Hurwitz stability criterion. But

$$J = \frac{1}{2\pi i} \int_{-i\infty}^{+i\infty} \frac{F(-s)}{A(s)} \, ds + \frac{1}{2\pi i} \int_{-i\infty}^{+i\infty} \frac{F(s)}{A(-s)} \, ds.$$

By a simple change of variable it follows that both of these integrals are equal. Hence we find that

$$J = \frac{f_1}{a_0}. \tag{1.37}$$

The evaluation of the Parseval integral has reduced to the problem of finding the coefficient f_1, given that $A(s)$ has all its roots in the negative half-plane. A proof of the Routh–Hurwitz stability criterion follows from the requirement that J is positive for arbitrary choices of the coefficients b_i of the numerator polynomial $B(s)$ (Walton and Gorecki, 1984).

Tabulated expressions for J_n, where n signifies the order of the $A(s)$ polynomial, are to be found in James, Nichols and Phillips (1947), Newton, Gould and Kaiser (1957), and Naslin (1968). Alternative methods are given in Åström (1970).

The extension of the methods of this section to other (symmetrical) criteria is straightforward. For example,

$$\int_0^\infty \left(\frac{de}{dt}\right)^2 dt \text{ and } \int_0^\infty \left(\frac{d^2e}{dt^2}\right)^2 dt$$

may be found by replacing $E(s)$, the Laplace transform of error, by the Laplace transform of corresponding derivative. Similarly, if even powers of t appear in the functional, e.g.

$$\int_0^\infty t^2 e^2(t) \, dt,$$

we may use the Laplace transform for $te(t)$ replacing the corresponding $E(s)$. In the cases mentioned so far, the symmetry is not broken, and the method applies directly. In the case of

$$\int_0^\infty t e^2(t) \, dt,$$

it is necessary to use the product form of Parseval's theorem

$$\int_0^\infty f(t)g(t) \, dt = \frac{1}{2\pi i} \int_{-i\infty}^{i\infty} F(s)G(-s) \, ds, \qquad (1.38)$$

and this is possible.

1.6 EVALUATION OF PERFORMANCE CRITERIA: MATRIX METHODS

We give here a brief introduction to Lyapunov techniques. One such technique (MacFarlane, 1963), having the merit of extending to other criteria as we show, is quoted at length in Jury (1974).

Suppose that

$$J = \int_0^\infty e^T(t) V e(t) \, dt, \qquad (1.39)$$

12 Performance criteria and delay-free systems [Ch. 1

where $e(t)$ is a column vector satisfying $\dot{e}(t) = Ae(t)$, A a constant matrix, and $e(0)$ given. $e^T(t)$ denotes the corresponding transpose and V is a positive definite symmetric matrix, possibly diagonal.

Let

$$\frac{d}{dt}[e^T(t)Pe(t)] = -e^T(t)Ve(t), \tag{1.40}$$

so that

$$\int_0^\infty e^T(t)Ve(t)\,dt = -\int_0^\infty \frac{d}{dt}[e^T(t)Pe(t)]\,dt = e^T(0)Pe(0), \tag{1.41}$$

where we suppose

$$\lim_{t\to\infty} e(t) = 0.$$

Now

$$\frac{d}{dt}[e^T(t)Pe(t)] = \frac{de^T(t)}{dt}Pe(t) + e^T(t)P\frac{d}{dt}e(t)$$

$$= e^T(t)A^TPe(t) + e^T(t)PAe(t)$$

$$= e^T(t)(A^TP + PA)e(t) = -e^T(t)Ve(t). \tag{1.42}$$

Hence solving the Lyapunov equation $A^TP + PA = -V$ for P, which for stability is known from section 1.3 to be positive definite, gives the cost $e^T(0)Pe(0)$.

Further, consider

$$\frac{d}{dt}\{te^T(t)Pe(t)\} = e^T(t)Pe(t) + t\frac{d}{dt}\{e^T(t)Pe(t)\}$$

$$= e^T(t)Pe(t) - t\{e^T(t)Ve(t)\}. \tag{1.43}$$

Integrating

$$\int_0^\infty \frac{d}{dt}\{te^T(t)Pe(t)\}\,dt = -[te^T(t)Pe(t)]_0^\infty = 0$$

$$= \int_0^\infty e(t)Pe(t)\,dt - \int_0^\infty te^T(t)Ve(t)\,dt.$$

$$\therefore \int_0^\infty te^T(t)Ve(t)\,dt = \int_0^\infty e(t)Pe(t)\,dt. \tag{1.44}$$

Sec. 1.6]   **Evaluation of performance criteria: matrix methods**   13

But the RHS is of the same form as the earlier problem and is equal to $e^T(0)P_1 e(0)$, where P_1 satisfies $A^T P_1 + P_1 A = -P$. Again the stability condition implies that P_1 is also positive definite.

Hence

$$\int_0^\infty t e(t)^T V e(t) \, dt = e(0)^T P_1 e(0). \qquad (1.45)$$

This process may be continued for higher powers of t in the performance integral.

A similar technique enables evaluations of integrals of the type

$$\int_0^\infty t^n e^{(n)}(t)^T V e^{(n)}(t) \, dt \qquad (1.46)$$

also, where $e^{(n)}(t)$ denotes the nth time derivative of $e(t)$.

The following technique is also a Lyapunov method, which makes use of the fundamental matrix $\Phi(t)$. It has the advantage of extending, with some modification, to the time-delay case.

With V as before we find

$$J = \int_0^\infty e^T(t) V e(t) \, dt. \qquad (1.47)$$

$\Phi(t)$ is the solution of

$$\frac{d\Phi}{dt} = A\Phi \text{ with } \Phi(0) = I_n.$$

The solution is the same as that of MacFarlane's method, as expected, but the method is instructive for later purposes. Writing $e(t) = \Phi(t)e(0)$ we have, denoting $e(0)$ by e_0

$$J = \int_0^\infty e^T(t) V e(t) \, dt$$

$$= \int_0^\infty e_0^T \Phi^T(t) V \Phi(t) e_0 \, dt$$

$$= e_0^T \int_0^\infty \Phi^T(t) V \Phi(t) \, dt \, e_0 = e_0^T M e_0, \text{ say.} \qquad (1.48)$$

If we introduce the function $M(s)$ defined by:

14   Performance criteria and delay-free systems   [Ch. 1

$$M(s) \equiv \int_s^\infty \Phi^T(t-s) V \Phi(t-s) \, dt \qquad (1.49)$$

it follows by change of variable that

$$M(s) = \int_0^\infty \Phi^T(t) V \Phi(t) \, dt \qquad (1.50)$$

and so $M(s)$ is constant and equal to M.

Hence, using the technique, differentiation under the integral sign

$$\frac{dM(s)}{ds} = 0 = \int_s^\infty \frac{\partial \Phi(t-s)^T}{\partial s} V \Phi(t-s) \, dt$$

$$+ \int_s^\infty \Phi(t-s)^T V \frac{\partial \Phi}{\partial s}(t-s) \, dt - V. \qquad (1.51)$$

Using

$$-\frac{\partial}{\partial s} \Phi(t-s) = \frac{\partial \Phi}{\partial t}(t-s) = A\Phi(t-s) \qquad (1.52)$$

and the commutativity of $A\Phi$, we obtain

$$A^T M + MA + V = 0. \qquad (1.53)$$

The solution of this Lyapunov equation, on which there is an extensive literature, is discussed in Barnett (1984), where many further references are to be found.

1.7 REVIEW OF THE LATER CHAPTERS

Time-delay systems are introduced in Chapter 2, and the properties that distinguish the behaviour of such systems from their delay-free counterparts are emphasized. A stability analysis is an obvious prerequisite to finding cost functionals. The stability analysis of time-delay systems has its distinctive problems, in particular that the number of system poles is in general infinite. A direct method for the analysis of stability with respect to the time-delay value is presented in this chapter arising from work recently published (Walton and Marshall, 1987a).

Successive chapters extend the techniques for cost function evaluation introduced in Chapter 1 for the delay-free case. Firstly in Chapter 3 the Heaviside expansion of error is squared and integrated for the case of an infinite number of poles, by explicit summations in the time-domain. These results are extended to other criteria of performance.

Chapter 4 exploits the Parseval relationship of the Laplace transform, and techniques are introduced for evaluating the resulting integral in the complex plane. In particular it is shown that there is a very close relationship with the direct stability method of Chapter 2. The reduction of the integral to summations of a finite number of residues is the central theme of this chapter, together with the extensions to systems with delays related by integers. Again the extensions to other criteria are explained.

Chapter 5 shows how matrix methods may be extended to the delay case. The evaluation of quadratic and other cost functionals is converted to two-point boundary value problems, and also the effects of non-zero initial conditions receive explicit treatment.

Chapter 6 examines implication for the discrete case, and shows that the techniques of evaluating Parseval integrals by reducing the number of residues are applicable here also.

Chapter 7 uses the results of the earlier chapters to quantify the accuracy of some delay-free techniques based on the use of the Padé approximant. Opportunity is taken also of examining the use of the Padé approximant in stability studies of time-delay systems.

Chapters 8, 9 and 10 explore the application of the techniques, and some evaluations made in the earlier chapters to problem of three-term control, and predictive methods of control. In particular the effect of mismatch in performance of predictive control receives detailed treatment.

2

Time-delay systems and stability

2.1 INTRODUCTION

For all physical systems there will be a duration of time, possibly very short, between the application of a stimulus, and the detection of the corresponding response. In some process plants the duration of the delay between stimulus and onset of response may be minutes, or even longer. In biological systems, delays may well be of a few hundred milliseconds duration (response-time in man), whereas in signal processing, delays measured in microseconds, or much shorter times, may be very important. The physical limit to speed of processing in digital computing systems is set by transition times, that is the time for electrons to travel finite distances. The delay between the transmission of an electromagnetic wave from an aerial and the reception of its reflection from a distant object though of short duration is the principal feature upon which radar is based. A more recent application is the use of delayed ultrasonic reflections for automatic focussing of cameras.

A pure time delay, an essential element in the modelling and description of these systems, has the property that input and output are identical in form, the only difference being that of translation along the time axis. The stimulus $x(t)$ results in a response $y(t)$ equal to $x(t - h)$ where h is the delay. In many complex systems there will be more than one delay, and these delays may be, for example, in measurement at an output, in control at an input or in a feedback path.

We shall consider, initially, systems which have a single delay. Input–output behaviour of single-delay systems and the associated stability properties will be explored in the context of single-input single-output systems. Later parts of the chapter will introduce the multivariable versions of such systems and systems with two or more delays will be analysed with respect to stability. Delays which are related by integers will be called commensurate delays, and delays not so related, general delays. In this chapter only those aspects of time-delay systems which are different in character from delay-free systems will be dwelt upon. The principal theme is that of distribution of poles, in closed-loop systems, and ways of checking whether such poles lie in the positive half of the complex plane—which characterizes instability.

2.2 SYSTEMS WITH A SINGLE TIME DELAY

Consider the solution of the delay-differential equation

$$\dot{y}(t) = -y(t - h) + H(t). \tag{2.1}$$

We assume that this describes a system quiescent at $t < 0$, the only stimulus being the step $H(t)$ applied at $t = 0$. h is the time delay. Fig. 2.1 shows a system to which equation (2.1) applies.

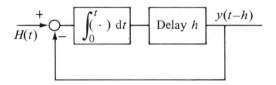

Fig. 2.1. System with equation: $\dot{y}(t) = -y(t - h) + H(t)$.

Before $t = h$, the delay in the feedback means that the derivative term is unaffected by the output so that $\dot{y}(t) = H(t)$ and $y(t) = tH(t)$.

At $t = h$ this time function will be 'stored in the delay' so to speak. In the interval $h < t < 2h$ this stored signal affects the derivative in addition to the effect of the input, so that when $h < t < 2h$ the equation to solve is

$$\dot{y}(t) = H(t) - (t - h)H(t - h). \tag{2.2}$$

In a sense the stored function acts as an additional input, and the problem is similar to that in the first time interval. Successive intervals are treated similarly with the response of the previous interval serving as input. In this way we are led to a succession of very similar problems which result in the solution

$$y(t) = tH(t) - \tfrac{1}{2}(t - h)^2 H(t - h) + \tfrac{1}{6}(t - 2h)^3 H(t - 2h) + \ldots$$
$$+ \frac{(-1)^r}{(r + 1)!}(t - rh)^{r+1} H(t - rh) + \ldots . \tag{2.3}$$

The method just described, 'the method of steps', extends in a natural way to higher-order systems, the only additional difficulties arising from the initial conditions on the differential coefficients at the start of each 'epoch' or step.

In the next section, we show that a quicker method of obtaining the solution is by the use of the Laplace transform. However, we first note that the transfer function

18 Time-delay systems and stability [Ch. 2

of the pure delay, h, is $\exp(-sh)$, an obvious consequence of the delay theorem of the Laplace transform

$$\mathscr{L}\{x(t-h)H(t-h)\} = X(s)\exp(-sh)$$

where

$$X(s) = \mathscr{L}\{x(t)H(t)\}. \tag{2.4}$$

2.3 METHOD OF STEPS IN THE TRANSFORM DOMAIN

Taking the Laplace transform of equation (2.1) results in

$$sY(s) + \exp(-sh)Y(s) = \frac{1}{s}. \tag{2.5}$$

Hence

$$Y(s) = \mathscr{L}\{y(t)\}$$

$$= \frac{1}{s}\frac{1}{s+\exp(-sh)}$$

$$= \frac{1}{s^2}\frac{1}{1+\frac{1}{s}\exp(-sh)} \tag{2.6}$$

which may be expanded to give

$$Y(s) = \frac{1}{s^2} - \frac{1}{s^3}\exp(-sh) + \frac{1}{s^4}\exp(-2sh)\ldots$$

$$= \mathscr{L}\left\{tH(t) - \frac{(t-h)^2}{2!}H(t-h) + \frac{(t-2h)^3}{3!}H(t-2h)\ldots\right\} \tag{2.7}$$

Whence

$$y(t) = \sum_{r=0}^{\infty}(-)^r\frac{(t-rh)^{r+1}}{(r+1)!}H(t-rh), \tag{2.8}$$

in agreement with (2.3).

For the more general system represented by Fig. 2.2, for which

$$Y(s) = \frac{G_1}{1+G_1G_2\exp(-sh)}X(s), \tag{2.9}$$

a similar calculation gives

$$Y(s) = X(s)G_1\sum_{r=0}^{\infty}(-)^r(G_1G_2)^r\exp(-srh) \tag{2.10}$$

and hence

$$y(t) = \sum_{r=0}^{\infty} (-)^r U_r(t - rh) H(t - rh) \tag{2.11}$$

where $U_r(t) = \mathscr{L}^{-1}\{G_1(G_1 G_2)^r\}$.

It is unusual for such summations to reduce to closed-form time-domain expressions.

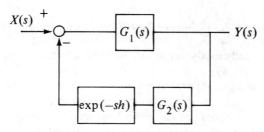

Fig. 2.2. System with delay in the feedback path.

For the system represented by Fig. 2.2, and other systems with a single delay, the corresponding error transform, $E(s)$, can be written, after rationalizing, as

$$E(s) = \frac{B(s) + D(s)\exp(-sh)}{A(s) + C(s)\exp(-sh)} \tag{2.12}$$

where $A(s)$, $B(s)$, $C(s)$, $D(s)$ are polynomials in s. We shall use equation (2.12) as the standard form for error in the single-delay case. In practical situations it is usual for the degree of $A(s)$ to be greater than the degrees of $B(s)$, $C(s)$, $D(s)$.

The corresponding characteristic equation of this standard form is

$$F(s, h) \equiv A(s) + C(s)\exp(-sh) = 0. \tag{2.13}$$

Specific examples giving rise to this characteristic equation are a delay-free system $C(s)/A(s)$ with delayed unity negative feedback (i.e. $G_1(s) = C(s)/A(s)$, $G_2(s) = 1$ in Fig. 2.2) and a system with forward transfer function $C(s)\exp(-sh)/A(s)$ with unity negative feedback.

Let the degree of $A(s)$ be n, and that of $C(s)$, m. It is standard to refer to the cases: $n > m$, $n = m$, $n < m$ as retarded, neutral, and advanced respectively. We note that the retarded case corresponds to the open-loop system, in examples, such as we have just mentioned, having a gain which decreases to zero at infinite frequencies, which will hold for physical systems.

2.4 STABILITY OF TIME-DELAY SYSTEMS WITH A SINGLE DELAY

Time-domain expressions (such as (2.11)) do not lead to stability criteria. For those, we demand that the system poles, i.e. the roots of the characteristic equation, have negative real parts.

By way of introduction we consider the characteristic equation corresponding to (2.5), namely $s + \exp(-sh) = 0$. Writing $s = \sigma + i\omega$ and equating real and imaginary parts leads to the pair of equations

$$\left.\begin{array}{l} \sigma + \exp(-\sigma h)\cos\omega h = 0 \\ \omega - \exp(-\sigma h)\sin\omega h = 0 \end{array}\right\} \tag{2.14}$$

for which the number of roots is infinite. This example shows the most important new feature in time-delay systems, i.e. that the number of system poles is infinite. Elimination of the trigonometric terms leads to $|s|^2 = \exp(-2\sigma h)$, from which it is concluded that for sufficiently large $|s|$, σ is negative, i.e. roots in left half-plane. Further, roots must satisfy $\tan\omega h = -\omega/\sigma$ (by elimination of the exponential terms) so that the large roots are separated (in ω) by $2\pi/h$ approximately.

Secondly, if $s = \sigma + i\omega$ is a solution of the characteristic equation then so is $s = \sigma - i\omega$ and so the roots are either real or occur in complex conjugate pairs, as will be the case for any characteristic equation with real coefficients and parameters.

As h varies, the roots will move and the question of asymptotic stability is that of determining for what (positive) values of h, if any, not all the roots lie in the left half-plane. A simpler problem to solve is that of determining for what values of h, if any, there are roots actually on the imaginary axis. With $\sigma = 0$, (2.14) reduce to $\cos\omega h = 0$ and $\sin\omega h = \omega$ which yield $\omega = \pm 1$, $\cos h = 0$, $\sin h = 1$. Thus when $h = \pi/2 + 2\pi q$ there are roots on the imaginary axis at $s = \pm i$, and these may or may not move, when h is increased, into the right half-plane. One way to answer the stability question is therefore that of developing a systematic procedure to discuss the behaviour of the roots as h increases. One such procedure, that of Walton and Marshall (1987a), will be given in detail below.

Other ways of looking at the question of stability include the graphical methods of Satché and Mikhailov, which are discussed in Popov (1962), Krall (1967) and Marshall (1979). Several analytical methods of varying complexity are offered by Pontryagin (1955), Hertz, Jury and Zeheb (1984), Thowsen (1981), Neimark (1949), Gorecki, Fuksa, Grabowski and Korytowski (1989), MacDonald (1989) and Stépán (1989).

As stated above, the condition for asymptotic stability is that all the roots of the characteristic equation

$$F(s, h) \equiv A(s) + C(s)\exp(-sh) = 0 \tag{2.15}$$

lie in the left half of the complex s-plane and the basic problem of stability is that of determining the range(s) of values of h for which this is so. The procedure introduced in Walton and Marshall (1987a) consists of three steps. The first step is to examine stability for $h = 0$ and determine the number of zeros, if any, of $F(s, 0)$ not lying in the left half-plane. The second step considers the case of infinitesimally small positive h when there will be an infinite number of new zeros and it is necessary to know whereabouts in the complex plane they have appeared. The third step is to find positive values of h, if any, at which there are zeros of (2.15) lying on the imaginary axis, i.e. find those values of ω and h for which $F(i\omega, h) = 0$, and then to determine whether the zeros merely touch the axis or whether they cross from one half-plane to the other with increasing h. Zeros crossing from left to right are called destabilizing and those from right to left, stabilizing.

Using the above procedure, starting with $h = 0$ and considering increasing h, it will be possible to study the movement of the zeros and, in particular, to determine when they do not all lie in the left half-plane, i.e. the regions of instability. As we will show, certain general results may be obtained, the implications of which are that for any given system whose characteristic equation is of the type given in (2.15) it is not necessary to do steps two and three *ab initio* but merely to consider a particular polynomial wich contains all the relevant information.

We now consider each of the three steps in detail, starting with the first step, which examines stability at $h = 0$ when (2.15) reduces to

$$F(s, 0) \equiv A(s) + C(s) = 0. \tag{2.16}$$

This is a delay-free problem to which any of the classical methods, Routh–Hurwitz for example, may be applied. If (2.16) is found to be unstable it will then be necessary to know how many zeros lie in the right half-plane or on the imaginary axis. In most practical examples this can be done by inspection. Routh's Algorithm (Gantmacher, 1964) may also be of use.

As stated above, one novel feature of time-delay systems is that there is an infinite number of roots of the corresponding characteristic equation. Consequently as h changes from being zero to being infinitesimally small and positive, the number of roots will change from being finite to being infinite. Step two addresses the question of whereabouts in the complex plane does this infinite number of *new* roots appear. As h is infinitesimally small the new roots must come in 'at infinity' otherwise $\exp(-sh)$ would be approximately equal to unity and they would not be new roots. Consequently, if the degree of $A(s)$ is greater than the degree of $C(s)$, (2.15) can be satisfied for large s if and only if $\exp(-sh)$ is large, i.e. Re $s < 0$. Thus for retarded systems the new roots all lie in the left half-plane. A similar argument shows that for advanced systems the new roots lie in the right half-plane. The case of neutral systems requires a more detailed investigation, and this will be done later in this chapter.

Step three considers potential crossing points on the imaginary axis. Firstly, it follows by taking complex conjugates that if $F(s, h) = 0$ has a root at $s = i\omega$ then it also has a root at $s = -i\omega$. Consequently if the roots cross or touch the imaginary axis they do so in pairs and it therefore suffices to consider positive ω. The exception to this is $s = 0$ and this special case will be discussed in more detail shortly.

We begin by seeking solutions of (2.15) of the form $s = \pm i\omega$. That is

$$\left. \begin{array}{l} A(i\omega) + C(i\omega) \exp(-i\omega h) = 0 \\ A(-i\omega) + C(-i\omega) \exp(i\omega h) = 0. \end{array} \right\} \tag{2.17}$$

Elimination of the exponential terms, which is equivalent to the elimination of h, yields

$$A(i\omega)A(-i\omega) - C(i\omega)C(-i\omega) = 0. \tag{2.18}$$

The expression on the left-hand side of this equation is a polynomial in ω^2 which, for retarded systems, will have the same degree as $A(s)$. For convenience, we denote this polynomial by $W(\omega^2)$, i.e.

$$W(\omega^2) \equiv A(i\omega)A(-i\omega) - C(i\omega)C(-i\omega). \tag{2.19}$$

Clearly only non-negative zeros of $W(\omega^2)$ are of interest since only these can lead to potential crossing points $s = \pm i\omega$.

One immediate consequence is that if there are *no* positive roots of $W(\omega^2) = 0$, as is the case, for example, when all the coefficients of powers of ω^2 in $W(\omega^2)$ are positive, there are no values of h for which $F(i\omega, h) = 0$. Consequently for such systems, assuming they are retarded, there will be no change in stability, i.e. if the system is stable at $h = 0$ it will be stable for all $h \geqslant 0$ whereas if it is unstable at $h = 0$ it will be unstable for all $h \geqslant 0$.

We have shown that if $F(s, h) = 0$ for $s = i\omega$ then $W(\omega^2) = 0$. However, if we have a real value of ω satisfying $W(\omega^2) = 0$ does it necessarily follow that there is some corresponding real positive value of h such that $F(i\omega, h) = 0$, i.e. such that

$$A(i\omega) + C(i\omega) \exp(-i\omega h) = 0. \tag{2.20}$$

This is equivalent to

$$\exp(-i\omega h) = -\frac{A(i\omega)}{C(i\omega)}. \tag{2.21}$$

We note that $C(i\omega) \neq 0$ since if this were not so, from (2.18), $A(i\omega)$ would also be zero, which is not possible since we naturally assume that any common factors of $A(s)$ and $C(s)$ have been removed.

Equation (2.21) will yield real positive values of h if and only if the modulus of the right-hand side is unity and, from (2.18) it follows that this is indeed the case. Thus for any real $\omega \neq 0$ satisfying $W(\omega^2) = 0$ there exist real positive h such that $F(i\omega, h) = 0$ and these are given by

$$\cos \omega h = \operatorname{Re}\left\{-\frac{A(i\omega)}{C(i\omega)}\right\}, \quad \sin \omega h = \operatorname{Im}\left\{\frac{A(i\omega)}{C(i\omega)}\right\}. \tag{2.22}$$

Thus if $h_0(\omega)$ denotes the smallest (positive) value of h, at a particular value of ω, satisfying this, then $h = h_0(\omega) + 2\pi q/\omega$, $q = 0, 1, 2, \ldots$, are solutions. Hence to each ω satisfying $W(\omega^2) = 0$ there are an infinite number of values of h.

As stated above, the case $s = 0$ (i.e. $\omega = 0$) requires special attention since

$$W(0) = 0 \Leftrightarrow A^2(0) - C^2(0) = 0$$
$$\Leftrightarrow A(0) + C(0) = 0 \text{ or } A(0) - C(0) = 0. \tag{2.23}$$

Now $A(0) + C(0) = 0$ implies that $s = 0$ is a solution for all values of h, i.e. $F(0, h) = 0$, and so the system is always unstable. On the other hand, if $A(0) - C(0) = 0$, we would require from (2.21), $\exp(-i\omega h) = -1$. Since $\omega = 0$ the only way this could occur is if h is infinite. We may therefore interpret such a solution as a point to which the roots tend as $h \to \infty$ and so for the purposes of studying crossing points they may be ignored.

Once we have found a particular value of h at which there is a root of the characteristic equation (2.15) lying on the imaginary axis, we must investigate its behaviour before and after, i.e. for slightly smaller and slightly larger values of h. In other words, as h varies, does the root cross the axis, if so in which direction, or merely touch the axis? If we regard the characteristic equation $F(s, h) = 0$ as determining s as a function of h and consider Re (ds/dh) evaluated at the appropriate $s = i\omega$, a positive value will correspond to the root crossing the imaginary axis from left to right (i.e. destabilizing) and a negative value one from right to left (i.e. stabilizing). If the value is zero it is necessary to consider higher-order derivatives.

Differentiating the characteristic equation (2.15) with respect to h we obtain

$$\frac{ds}{dh} = \frac{C(s)s \exp(-sh)}{A'(s) + C'(s) \exp(-sh) - C(s)h \exp(-sh)} \tag{2.24}$$

where primes, as always, denote differentiation with respect to the argument of the function, which in this case is the variable s. With the aid of (2.15) this may be written

$$\frac{ds}{dh} = -s\left[\frac{A'(s)}{A(s)} - \frac{C'(s)}{C(s)} + h\right]^{-1}. \tag{2.25}$$

We are, in fact, interested in only the sign of the real part of this. So at $s = i\omega$ where $W(\omega^2) = 0$,

$$S \equiv \operatorname{sgn} \operatorname{Re} \frac{ds}{dh} = -\operatorname{sgn} \operatorname{Re} i\omega \left[\frac{A'(i\omega)}{A(i\omega)} - \frac{C'(i\omega)}{C(i\omega)} + h\right]^{-1}$$

$$= -\operatorname{sgn} \operatorname{Re} \frac{1}{i\omega}\left[\frac{A'(i\omega)}{A(i\omega)} - \frac{C'(i\omega)}{C(i\omega)} + h\right] \tag{2.26}$$

since the sign of the real part of a complex number is equal to the sign of the real part of its inverse. Thus

$$S = \operatorname{sgn} \operatorname{Im} \frac{1}{\omega}\left[\frac{C'(i\omega)}{C(i\omega)} - \frac{A'(i\omega)}{A(i\omega)}\right] \tag{2.27}$$

which we note to be independent of h. Thus although there are an infinite number of values of h associated with each particular value ω, the behaviour of the roots at these points will always be the same. Hence we may classify solutions of $W(\omega^2) = 0$ as stabilizing or destablizing according to whether S is negative or positive respectively.

Now

$$A(i\omega)A(-i\omega) = C(i\omega)C(-i\omega) > 0 \tag{2.28}$$

since A and C cannot be simultaneously zero and so (2.27) may be written

$$S = \operatorname{sgn} \operatorname{Im} \frac{1}{\omega}\left[C'(i\omega)C(-i\omega) - A'(i\omega)A(-i\omega)\right]$$

$$S = \text{sgn}\, \frac{1}{2i\omega} \{C'(i\omega)C(-i\omega) - C(i\omega)C'(-i\omega)$$
$$- A'(i\omega)A(-i\omega) + A(i\omega)A'(-i\omega)\}$$
$$S = \text{sgn}\, W'(\omega^2) \tag{2.29}$$

in which the prime denotes in all cases differentiation with respect to the argument. Thus provided S exists, the simple criterion (2.29) can be used to determine whether a root is stabilizing or destabilizing.

The case $S = 0$ (i.e. $W'(\omega^2) = 0$) requires further analysis since in such a case the root may cross the axis or merely touch it. In Walton and Marshall (1987a) the above result is generalized and it is shown that the behaviour of $W(\omega^2)$ near a zero exactly mirrors that of the corresponding root of $F(s, h) = 0$. In other words, if $W(\omega^2)$ crosses the axis at a particular value of ω then the corresponding root of $F(s, h) = 0$ crosses the imaginary axis at $s = i\omega$ whereas if $W(\omega^2)$ only *touches* the axis then the corresponding root of $F(s, h) = 0$ only *touches* the imaginary axis.

As can ben seen from the above, a study of the polynomial $W(\omega^2)$ will yield a considerable amount of information about the stability of the system in question since the positive zeros give the potential crossing points and the behaviour of $W(\omega^2)$ in the neighbourhood of these zeros gives the nature of the potential crossing points. The polynomial $W(\omega^2)$ also yields further information which is of particular importance for neutral systems.

Earlier it was stated that for infinitesimally small positive h, an infinite number of new roots of $F(s, h) = 0$ appears at infinity and that these are in the left half-plane for retarded systems and in the right half-plane for advanced systems. Now if

$$\left. \begin{array}{l} A(s) = a_0 s^n + a_1 s^{n-1} + \ldots + a_n \\ C(s) = c_0 s^m + c_1 s^{m-1} + \ldots + c_m \end{array} \right\} \tag{2.30}$$

where $a_0, c_0 \neq 0$, then for retarded systems $n > m$ and so for large ω, $W(\omega^2) \approx a_0^2 \omega^{2n}$. On the other hand, for advanced systems $n < m$ and so for large ω, $W(\omega^2) \approx -c_0^2 \omega^{2m}$. Thus for retarded systems, in which these new roots are in the left half-plane (i.e. are stable), W is positive for large ω whereas for advanced systems, in which the new roots are in the right half-plane (i.e. are unstable), W is negative for large ω.

The case of neutral systems is somewhat more complicated since, in this case, it is possible for all the roots of the characteristic equation to lie in the left half of the complex plane but for the system to be unstable. For example, the new roots of $s + 1 + se^{-sh} = 0$ that occur for infinitesimally small h have large $|s|$ so are given approximately by $1 + e^{-sh} = 0$, the solutions of which lie on the imaginary axis. A detailed calculation will show that the roots are just in the left half-plane but that the system is not stable.

For our purposes the following statements are sufficient:

(a) A neutral system is (asymptotically) stable if there exists a real $d < 0$ such that all the solutions of the characteristic equation $F(s, h) = 0$ have $\text{Re}\, s < d$.
(b) A neutral system is not (asymptotically) stable if there exists a solution of the characteristic equation with $\text{Re}\, s > 0$.

Consequently it is step two in the above procedure that requires further consideration. Thus if $s = \sigma + i\omega$ is a new root of $F(s, h) = 0$ for infinitesimally small h, then $|\omega| \gg |\sigma|$ and

$$\exp(-h\sigma) = \left|\frac{A(s)}{C(s)}\right| \simeq \left|\frac{A(i\omega)}{C(i\omega)}\right| \tag{2.31}$$

from which we conclude that $\sigma > 0$ if and only if $|A(i\omega)| < |C(i\omega)|$, i.e. $W(\gamma^2) < 0$ for large ω. We may therefore conclude that the system is unstable if $W(\omega^2) < 0$ for large ω.

On the other hand, if $W(\omega^2) > 0$ for large ω the new roots will lie in the left half-plane but this is insufficient for stability. What is required is that they lie to the left of the line $\text{Re}\, s = d$, for some real $d < 0$. From (2.31), this will be possible if $\lim_{\omega \to \infty} |A(i\omega)/C(i\omega)| > 1$, i.e. if $|a_0/c_0| > 1$. Thus, if $W(\omega^2) > 0$ for large ω and $|a_0| > |c_0|$ then the system is stable for infinitesimally small h.

Before considering some examples we summarixe the results obtained so far:

Step One: Examine the stability at $h = 0$.
Step Two: Consider infinitesimally small positive h.

If the system is retarded, all the new roots will lie in the left half-plane and this step may be omitted. If the system is advanced then the infinite number of new roots all appear in the right half-plane and since they can cross into the left half-plane at only a finite rate for all finite values of h, there will be roots in the right half-plane and so the system will be unstable for all $h > 0$, although it is possible that such systems are stable at $h = 0$. Consequently step three is not needed for advanced systems. For neutral systems the location of the new roots is determined by the sign of $W(\omega^2)$ for large ω.

Step Three: Determine the positive zeros of the polymomial $W(\omega^2)$, the corresponding positive values of h and the nature of these zeros.

We remark that there will only be a finite number of such zeros since $W(\omega^2)$ is of degree n for retarded systems and of degree no greater than n for neutral systems. The nature of a zero is determined by the behaviour of $W(\omega^2)$ near that point; however, we *emphasize* that it is *never* necessary to calculate $W'(\omega^2)$, as its sign may be inferred. If $W(\omega^2) = 0$ has no repeated roots the stabilizing and destabilizing roots alternate. Moreover those systems of interest have $W(\omega^2) > 0$ for large ω and so the highest root is always destabilizing. In particular if there is only one root it must be destabilizing. More generally the roots may be labelled in descending order as destabilizing, stabilizing, etc. and the corresponding values of h similarly labelled. In this way, it is possible to determine for what values of h all the roots of $F(s, h) = 0$ lie in the left half-plane. If there are repeated roots of $W(\omega^2) = 0$ these may correspond to crossing points or merely points at which the roots touch. Consequently the behaviour of the polynomial $W(\omega^2)$ is of paramount importance in the stability analysis and although it is not essential to sketch $W(\omega^2)$ it may be found helpful, especially for higher-order examples.

The examples that follow clarify the procedure and discuss extra points of detail. Also, unless otherwise stated, the examples chosen are of retarded systems and so it is not necessary to perform Step Two.

Example 2.1 $F(s, h) = s + K \exp(-sh)$, $K > 0$

(a) $F(s, 0) = s + K$
Thus the system is stable at $h = 0$.

(b) $A(s) = s$ and $C(s) = K$ and so

$$W(\omega^2) \equiv A(i\omega)A(-i\omega) - C(i\omega)C(-i\omega)$$
$$= \omega^2 - K^2.$$

Thus there is only one positive solution of $W(\omega^2) = 0$, namely $\omega^2 = K^2$, and hence is destabilizing. The corresponding values of h are given by (2.22), i.e.

$$\cos \omega h = \text{Re}\left\{-\frac{i\omega}{K}\right\} = 0, \quad \sin \omega h = \text{Im}\left\{\frac{i\omega}{K}\right\} = 1$$

and so

$$h = (4q + 1)\frac{\pi}{2K}, \quad q = 0, 1, 2, \ldots.$$

Thus at $h = \pi/2K$, two roots of $F(s, h) = 0$ cross from the left to the right (one at $s = iK$ and the other at $s = -iK$). Then two more cross at $h = 5\pi/2K$ and so on. We therefore conclude that the only region of stability is $0 \leqslant h < \pi/2K$.

Example 2.2 $F(s, h) = s^2 + 4s + 4 - (1/4) \exp(-sh)$

(a) $F(s, 0) = s^2 + 4s + 15/4$. Hence stability at $h = 0$.

(b) $W(\omega^2) = (-\omega^2 + 4i\omega + 4)(-\omega^2 - 4i\omega + 4) - 1/16$
$= (\omega^2 + 15/4)(\omega^2 + 17/4) = 0$.

Here, with no positive solutions for ω^2, we conclude stability for all $h \geqslant 0$, i.e. stable independent of delay.

Example 2.3 $F(s, h) = s^3 + s^2 + 2s + 1 + \exp(-sh)$

(a) $F(s, 0) = s^3 + s^2 + 2s + 2 = (s + 1)(s^2 + 2)$.

This has two roots on the imaginary axis, and one LH root.

(b) $W(\omega^2) = (-i\omega^3 - \omega^2 + 2i\omega + 1)(i\omega^3 - \omega^2 - 2i\omega + 1) - 1$
$= \omega^2(\omega^2 - 1)(\omega^2 - 2)$.

Although $\omega^2 = 0$ is a root of $W(\omega^2) = 0$, $A(0) = C(0) = 1$ and so $s = 0$ is *not* a solution of $F(s, h) = 0$ for finite h and so may be discarded. As mentioned earlier, the interpretation of this zero of $W(\omega^2)$ is that the roots of $F(s, h) = 0$ tend to $s = 0$ as

$h \to \infty$. The other zeros of $W(\omega^2)$ are $\omega^2 = 2$ and $\omega^2 = 1$. As $\omega^2 = 2$ is the larger of the two it is destabilizing and the corresponding values of h are given by

$$\cos \omega h = \text{Re}\{i\omega^3 + \omega^2 - 2i\omega - 1\} = 1$$

$$\sin \omega h = \text{Im}\{-i\omega^3 - \omega^2 + 2i\omega + 1\} = 0,$$

i.e. $h = q\pi\sqrt{2}$, $q = 0, 1, 2, \ldots$. Thus as h is increased, the two imaginary roots at $h = 0$ move into the right half-plane, giving instability. Secondly as $\omega^2 = 1$ is the smaller of the two zeros of $W(\omega^2)$ it is stabilizing and the corresponding values of h are given by

$$\cos \omega h = 0, \quad \sin \omega h = \omega,$$

i.e. $h = (2q + \tfrac{1}{2})\pi$, $q = 0, 1, 2, \ldots$. Thus at $h = 0, \pi\sqrt{2}, 2\pi\sqrt{2}, \ldots$ a pair of roots crosses from the left half-plane into the right half-plane whereas at $h = \pi/2, 5\pi/2, 9\pi/2, \ldots$ a pair of roots crosses from the right half-plane into the left half-plane. The timetable of events is therefore as follows. The two roots originally on the imaginary axis move into the right half-plane and then move back into the left half-plane at $h = \pi/2$. At $h = \pi\sqrt{2}$ a second pair of roots crosses into the right half-plane and then crosses back at $h = 5\pi/2$. At $h = 2\pi\sqrt{2}$ a third pair crosses into the right and then crosses back at $h = 9\pi/2$, and so on. The stability ranges are given by the values of h when there are no roots in the right half-plane and to determine these it is helpful to know the values of h at which there are crossings, as shown in Table 2.1. From this table it may be seen that there are two roots in the right half-plane for $0 < h < \pi/2$, none for $\pi/2 < h < \pi\sqrt{2}$, two for $\pi\sqrt{2} < h < 5\pi/2$, none for $5\pi/2 < h < 2\pi\sqrt{2}$, two for $2\pi\sqrt{2} < h < 3\pi\sqrt{2}$ then four for $3\pi\sqrt{2} < h < 9\pi/2$. This is because the fourth pair of roots crosses into the right before the third pair have crossed back into the left. This is because the larger value of ω corresponds to the destabilizing crossing and consequently the interval between such crossings (i.e. $2\pi/\omega$) will be smaller than that for the stabilizing crossings. Thus the roots will accumulate in the right half-plane. In particular, once there have been two consecutive destabilizing crossings, no more stability intervals are possible since there can never be two consecutive stabilizing crossings. We therefore conclude that in this example there is stability for $\pi/2 < h < \pi\sqrt{2}$ and for $5\pi/2 < h < 2\pi\sqrt{2}$. We say these are two 'stability windows' and the next section gives a more detailed study of this phenomenon.

Table 2.1. Critical values of delay

$\omega_1 = 1$	Stabilizing	—	0.5π	—	2.5π	—	—
$\omega_2 = \sqrt{2}$	Destabilizing	0	—	$\sqrt{2}\pi$	—	$2\sqrt{2}\pi$	$3\sqrt{2}\pi$

Our final remark is that this example demonstrates that the addition of delay *can* have a stabilizing effect, a result that may be contrary to one's intuition.

Example 2.4 $F(s, h) = s^2 + s + 1 + s\exp(-sh)$

(a) $F(s, 0) = s^2 + 2s + 1 = (s + 1)^2$: stable.

(b) $W(\omega^2) = (-\omega^2 + i\omega + 1)(-\omega^2 - i\omega + 1) - (i\omega)(-i\omega)$
$= (\omega^2 - 1)^2 = 0.$

Here there is a *double* root at $\omega^2 = 1$. Clearly $W(\omega^2)$ touches the horizontal axis once for positive ω^2. Hence the system is stable for all values of h except the point values satisfying $\cos h = -1$, $\sin h = 0$, i.e. $h = (2q + 1)\pi$, where $q = 0, 1, 2, \ldots$.

Example 2.5 $F(s, h) = s^3 + 2s^2 + 2s + (2s + 1)\exp(-sh)$

(a) $F(s, 0) = s^3 + 2s^2 + 4s + 1 = 0$, stable, Routh–Hurwitz $a_3 a_0 < a_1 a_2$.

(b) $W(\omega^2) = (-i\omega^3 - 2\omega^2 + 2i\omega)(+i\omega^3 - 2\omega^2 - 2i\omega) - (2i\omega + 1)(-2i\omega + 1)$
$= \omega^6 - 1 = 0$, so that $\omega^2 = 1$ is the only real root.

This will occur when $\cos h = 0$, $\sin h = 1$,

i.e. $h = (2q + 1/2)\pi$, $q = 0, 1, 2, \ldots$.

As there is only one destabilizing value of ω^2 we have stability for $0 \leq h < \pi/2$ only.

Example 2.6 $F(s, h) = s + 1 + (s + 2)\exp(-sh)$

We note that as $A(s) = s + 1$ and $C(s) = s + 2$ are both of degree one this is an example of a neutral system and it will therefore be necessary to consider Step Two, i.e. to consider $W(\omega^2)$ for large ω^2.

(a) $F(s, 0) = 2s + 3 = 0$, stable.

(b) $W(\omega^2) = (i\omega + 1)(-i\omega + 1) - (i\omega + 2)(-i\omega + 2) = -3.$

Thus as $W(\omega^2) < 0$ for large ω^2 the infinite number of new roots occurs in the right half-plane so the system is unstable for all $h > 0$. However, this system *is* stable for $h = 0$. This phenomenon is known as *practical stability*, and is absent from retarded systems, as there the loop gain tends to zero at high frequencies, thus making instability impossible for infinitesimally small h, when absent at $h = 0$.

Example 2.7 $F(s, h) = 2s + 1 + s\exp(-sh)$

Again both $A(s) = 2s + 1$ and $C(s) = s$ are of degree one so this is also an example of a neutral system and it will again be necessary to consider Step Two.

(a) $F(s, 0) = 3s + 1 = 0$, stable

(b) $W(\omega^2) = (2i\omega + 1)(-2i\omega + 1) - (i\omega)(-i\omega) = 3\omega^2 + 1.$

Thus as $W(\omega^2) > 0$ for large ω^2 the infinite number of new roots occurs in the left half-plane. Secondly, as $a_0 = 2$ and $c_0 = 1$, $|a_0/c_0| > 1$ so the system is stable for infinitesimally small positive h. Then as there are no positive solutions of $W(\omega^2) = 0$ there are no crossing points and hence the system is stable for all h.

2.5 STABILITY WINDOWS

To show that even low-order systems can exhibit quite complicated behaviour we give the following example and a note about the number of 'stability windows'. These should act as a cautionary note to any reader who might be attempting to solve equations of the single-delay type by using approximations, such as the Padé approximant, to analyse stability of time-delay systems.

Example 2.8
Consider the quadratic case
$$F(s, h, b) = s^2 + s + b^2 + b\exp(-sh) = 0, \, b > 1.$$
The system is stable at $h = 0$, for all positive b.
$$W(\omega^2) = \omega^4 + \omega^2(1 - 2b^2) + b^2(b^2 - 1) = 0, \text{ with roots } \omega_1^2 = b^2 - 1, \omega_2^2 = b^2.$$
Values of h must satisfy
$$b^2 - \omega^2 + i\omega + b\cos\omega h - ib\sin\omega h = 0 \text{ at these values of } \omega,$$
so that $b^2 - \omega^2 + b\cos\omega h = 0$

$\omega - b\sin\omega h = 0.$

At $\omega = \omega_1$: $b\cos\omega_1 h = -1$, $b\sin\omega_1 h = (b^2 - 1)^{1/2}$

whence

$$\omega_1 h = \pi - \cos^{-1}(1/b) + 2q\pi \text{ and } h_1(q) = ((2q+1)\pi - \cos^{-1}(1/b))/(b^2-1)^{1/2}.$$

At $\omega = \omega_2$: $\cos\omega_2 h = 0$, $\sin\omega_2 h = 1$, and hence $h_2(q) = (2q + \frac{1}{2})\pi/b$. As $\omega_2 > \omega_1$ the values of $h(= h_1(q))$ arising from ω_1 are 'stabilizing' values, and those arising from ω_2 are destabilizing. Clearly, $h_1(q)$ and $h_2(q)$ are increasing functions of q.

The system is stable at $h = 0$, the first destabilizing value is $h_2(0)$, the first stabilizing value is $h_1(0)$, and as long as the terms in the sequences $\{h_2(q)\}, \{h_1(q)\}$ continue to alternate there will be a succession of stable and unstable regions. This interlacing *must* eventually cease since $2\pi/\omega_1$, the interval between successive stabilizing values, is greater than $2\pi/\omega_2$, that of the destabilizing values. When it does, permanent instability will occur. The condition of a stability window to occur is that $h_2(q+1) - h_1(q) > 0$. Denoting by q^* the lowest integer value for which this inequality fails to hold we find that q^*, which is the number of stability windows, satisfies

$$q^* = \left[0.75 + b\left(b + \sqrt{(b^2-1)}\right)\left(0.75 + \frac{0.5}{\pi}\cos^{-1}\left(\frac{1}{b}\right)\right)\right]$$

where $[a]$ denotes greatest integer less than a, and $q^* \approx 2b^2$ for large b. For the case $b = \sqrt{2}$, $q^* = [3.737] = 3$. For $b = 20$, $q^* = 793$ ($2b^2 = 800$).

In higher-order cases, where there may be more than two positive roots in ω^2, it follows that the analysis of stability windows will be more complicated than that of the above example. However, a simple tabular display of increasing h values classified as stabilizing or destabilizing will quickly clarify the situation. There will be no

2.6 SYSTEMS WITH MORE THAN ONE DELAY: COMMENSURATE DELAYS

Consider the matrix delay-differential equation

$$\dot{x}(t) = A_0 x(t) + A_1 x(t - h) + Bu(t), \tag{2.32}$$

in which the state $x(t)$ is delayed. Stability analysis of such a system may be reduced to finding the roots of its characteristic equation

$$\det [sI - A_0 - A_1 \exp(-sh)] = \det [sI - A_0 - A_1 z] = 0. \tag{2.33}$$

It is convenient, here and subsequently, to replace $\exp(-sh)$ by z.

Consider the example

$$A_0 = \begin{bmatrix} a_{11} & a_{12} \\ a_{21} & a_{22} \end{bmatrix}, \quad A_1 = \begin{bmatrix} \alpha_{11} & \alpha_{22} \\ \alpha_{21} & \alpha_{22} \end{bmatrix}, \tag{2.34}$$

for which the characteristic equation is

$$F(s, h) = s^2 - s(a_{11} + a_{22} + \alpha_{11}z + \alpha_{22}z) + (a_{11} + \alpha_{11}z)(a_{22} + \alpha_{22}z)$$
$$- (a_{12} + \alpha_{12}z)(a_{21} + \alpha_{21}z) = 0. \tag{2.35}$$

This characteristic equation involves not only the single-delay term $z = \exp(-sh)$ but also a delay of $2h$ represented by $z^2 = \exp(-2sh)$. In fact we see that conditions on A_0, A_1 will need to be very restricting in order that either all the $\exp(-sh)$ terms are missing, or that all the $\exp(-2sh)$ terms are missing.

When A_0, A_1 are of order n then in general all delays from $\exp(-sh)$ to $\exp(-nsh)$ will be present. Delays such as these, related by integers, are called 'commensurate' delays.

The stability problem reduces to two single-delay cases when A_0 and A_1 commute and the characteristic equation then factorizes into terms of the form

$$(s + \beta_{0q} + \beta_{1q} \exp(-sh)), \quad q = 1, 2, \ldots, n.$$

Commutativity is, however, rare in practical systems.

Commensurate delays arise naturally in some transmission line problems (Grabowski, 1989). In a line which satisfies the distortionless condition but which is incorrectly terminated, reflections will occur at the receiving end. The distortionless line with a resistive termination not equal to the characteristic impedance, which is resistive in the distortionless case, has as its transfer function

$$G(s) = (\delta \sinh \gamma + \cosh \gamma)^{-1} \tag{2.36}$$

where $\gamma = \alpha + s\beta$, and δ is the ratio of characteristic to terminating impedance, in this case a real number. α and β are positive constants related to the line constants. Hence

Sec. 2.7] Stability test for commensurate delays

$$G(s) = \frac{2}{\delta(e^{\gamma} - e^{-\gamma}) + (e^{\gamma} + e^{-\gamma})} = \frac{2e^{-\gamma}}{(\delta + 1) + (1 - \delta)e^{-2\gamma}} \quad (2.37)$$

$e^{-\gamma} = e^{-\alpha}e^{-s\beta} = \kappa e^{-sh}$ say, an attenuation, and a pure delay. Hence

$$G(s) = \frac{2\kappa \exp(-sh)}{(\delta + 1) + (1 - \delta)\kappa^2 \exp(-2sh)}. \quad (2.38)$$

This is clearly an example where commensurate delays occur in a system. We note when $\delta = 1$, i.e. correct termination, that $G(s) = \kappa \exp(-sh)$, a single delay.

These examples lead to the stability problem for the commensurate case.

2.7 STABILITY TEST FOR COMMENSURATE DELAYS

The steps involved in the stability analysis of systems with commensurate delays follow those of the single-delay case. The first step is to examine stability for $h = 0$ and involves nothing new. The second step is to consider the infinite number of new zeros that occur for infinitesimally small positive h. For simplicity, we will consider only retarded systems, i.e. those for which the degree of the term independent of z is strictly greater than that of the coefficient of any power of z. Consequently all these new roots are in the left half-plane.

The third step is to find those values of h at which there are potential crossing points and then to determine whether or not, as h increases, the roots cross the imaginary axis and, if they do, in which direction. Before proceeding to the general case, an example with only the delays h and $2h$ is given in order to demonstrate the principles involved.

Example 2.9 $F(s, h) = s + \exp(-sh) + \exp(-2sh) = 0.$ (2.39)

(a) $F(s, 0) = s + 2 = 0$, stable.
(b) From the previous result for a single delay, we seek common roots of $F(s, h) = 0$, and $F(-s, h) = 0$ for $s = i\omega$. This leads to

$$s + z + z^2 = 0 \quad \text{and} \quad -s + \frac{1}{z} + \frac{1}{z^2} = 0$$

where $\exp(-sh)$ has again been denoted by z, i.e.

$$s + z + z^2 = 0 \quad \text{and} \quad -sz^2 + z + 1 = 0.$$

Eliminating the term in z^2 gives $(z + 1) = -s(z + s)$, i.e.

$$z(1 + s) + (1 + s^2) = 0.$$

Eliminating the term independent of z gives $z^2 + z = -s(sz^2 - z)$ which, on cancelling z, gives

$$z(1 + s^2) + (1 - s) = 0.$$

Thus

$$z(1 + s^2) + (1 - s) = 0 \brace z(1 + s) + (1 + s^2) = 0$$

hence, eliminating z,

$$(1 + s^2)^2 - (1 - s^2) = 0$$

which gives $s^2(s^2 + 3) = 0$ so $s = 0$ (not a solution of $F(s, h) = 0$), or $-s^2 = \omega^2 = 3$. Now

$$\frac{1}{z} = \exp(i\omega h) = -\frac{(1 + s^2)}{1 - s} = \frac{1}{2} + i\frac{\sqrt{3}}{2} = \exp(i\pi/3). \tag{2.40}$$

Hence

$$\sqrt{3}h = \frac{\pi}{3} + 2\pi q, \quad \text{and} \quad h(q) = \frac{\pi}{3\sqrt{3}} + \frac{2\pi q}{\sqrt{3}}.$$

In the single-delay case it was shown that for every value of ω obtained (with the possible exception of $\omega = 0$) there was a corresponding real positive value of h (in fact an infinite number of them) such that the original characteristic equation was satisfied. In the commensurate delay case this question is more complicated since in this example the values of h were determined not from the original characteristic equation (2.39) but from a resulting equation (2.40). Here it is straightforward to confirm that $s = i\sqrt{3}$, $h = h(q)$ satisfies the original characteristic equation (2.39); however, this question must be addressed for the general case.

As $s = \pm i\sqrt{3}$ are the only potential crossing points it is reasonable to assume that they are destabilizing and hence the system has one stability window $0 \leq h < \pi/3\sqrt{3}$. In fact, ds/dh at $s = \pm i\sqrt{3}$ may be shown to have a positive real part (independent of h) confirming the above assumption. However, as we shall show shortly, an explicit calculation of this derivative is not necessary in the general case since there again exists a polynomial $W(\omega^2)$ which, with a slight modification, yields a sensitivity result analogous to the earlier one which concerned the sign of $W'(\omega^2)$ at a solution of $W(\omega^2) = 0$.

We now consider the general case of multiple commensurate delays, the characteristic equation of which will be

$$F(s, h) \equiv \sum_{k=0}^{n} A_k(s) \exp(-ksh) = \sum_{k=0}^{n} A_k(s) z^k = 0. \tag{2.41}$$

Firstly there is the usual preliminary of examining the stability at $h = 0$ by standard methods. The next step would normally be that of examining stability at infinitesimally small positive h. However, here this will not be necessary since we assume that the polynomial $A_0(s)$ is of higher degree than any of the polynomials $A_1(s), A_2(s), \ldots, A_n(s)$.

The next step is the determination of any potential crossing points, i.e. we seek solutions of the form $s = i\omega$. We therefore seek solutions of

$$F(s, h) \equiv \sum_{k=0}^{n} A_k(s) z^k = 0 \tag{2.42}$$

and, replacing s by $-s$,

$$F(-s, h) \equiv \sum_{k=0}^{n} A_k(-s) z^{-k} = 0. \tag{2.43}$$

One could use the classical methods (Burnside and Panton, 1960) to find the common roots; however, the following method generalizes the single-delay case (and the technique of Example 2.9) and gives a procedure for the systematic reduction in degree by elimination of the highest power of z. This iterative scheme eventually yields an equation independent of z (or equivalently h), i.e. the usual $W(\omega^2) = 0$.

Define

$$\begin{aligned} F^{(1)}(s, h) &= A_0(-s) F(s, h) - A_n(s) z^n F(-s, h) \\ &= \sum_{k=0}^{n-1} \{A_0(-s) A_k(s) - A_n(s) A_{n-k}(-s)\} z^k, \end{aligned} \tag{2.44}$$

i.e. a polynomial in z of degree $n - 1$, and

$$\begin{aligned} F^{(1)}(-s, h) &= A_0(s) F(-s, h) - A_n(-s) z^{-n} F(s, h) \\ &= \sum_{k=0}^{n-1} \{A_0(s) A_k(-s) - A_n(-s) A_{n-k}(s)\} z^{-k}. \end{aligned} \tag{2.45}$$

Then if $s = i\omega_0$ is a common root of $F(s, h) = 0$ and $F(-s, h) = 0$ it is also a common root solution of $F^{(1)}(s, h) = 0$ and $F^{(1)}(-s, h) = 0$, i.e.

$$\left.\begin{aligned} F^{(1)}(s, h) &= \sum_{k=0}^{n-1} A_k^{(1)}(s) z^k = 0 \\ F^{(1)}(-s, h) &= \sum_{k=0}^{n-1} a_k^{(1)}(-s) z^{-k} = 0 \end{aligned}\right\} \tag{2.46}$$

where $A_k^{(1)}(s) = A_0(-s) A_k(s) - A_n(s) A_{n-k}(-s)$.

This procedure is now repeated with these equations, being of the same form as the original pair, and, in general, with

$$\begin{aligned} F^{(r+1)}(s, h) &= A_0^{(r)}(-s) F^{(r)}(s, h) - A_{n-r}^{(r)}(s) z^{n-r} F^{(r)}(-s, h) \\ &= \sum_{k=0}^{n-r-1} A_k^{(r+1)}(s) z^k \end{aligned} \tag{2.47}$$

where $A_k^{(r+1)}(s) = A_0^{(r)}(-s) A_k^{(r)}(s) - A_{n-r}^{(r)}(s) A_{n-k-r}^{(r)}(-s)$.

Repeated application leads to the result that if $s = i\omega_0$ is a solution of $F(s, h) = 0$, then $s = i\omega_0$ is also a solution of $F^{(n)}(s, h) = A_0^{(n)}(s) = 0$ where $A_0^{(n)}(s) = A_0^{(n-1)}(-s) A_0^{(n-1)}(s) - A_1^{(n-1)}(s) A^{(n-1)}(-s)$ and so $W(\omega_0^2) = A_0^{(n)}(i\omega_0)$.

The corresponding values of h are then found from the equation linear in z, i.e. $A_0^{(n-1)}(s) + A_1^{(n-1)}(s) z = 0$. Therefore

$$\left.\begin{aligned} \cos \omega_0 h &= \text{Re}\{-A_0^{(n-1)}(i\omega_0)/A_1^{(n-1)}(i\omega_0)\} \\ \sin \omega_0 h &= \text{Im}\{A_0^{(n-1)}(i\omega_0)/A_1^{(n-1)}(i\omega_0)\} \end{aligned}\right\}. \tag{2.48}$$

Exactly as in the single-delay case these equations will yield real positive values for h (cf. equations (2.22)) as $|A_0^{(n-1)}(i\omega_0)/A_1^{(n-1)}(i\omega_0)| = 1$ since $W(\omega_0^2) = 1$. However, there is also the question of whether or not all the positive solutions of $W(\omega_0^2) = 0$, together with the corresponding h, are solutions of the original characteristic equation.

We have shown that if $s = i\omega_0$ is a solution of

$$F(s, h) = 0 \tag{2.49}$$

for a given h then it is also a solution of

$$F(-s, h) = 0 \tag{2.50}$$

and of

$$F^{(1)}(s, h) \equiv A_0(-s)F(s, h) - A_n(s)z^n F(-s, h) = 0 \tag{2.51}$$

and

$$F^{(1)}(-s, h) \equiv A_0(s)F(-s, h) - A_n(-s)z^{-n} F(s, h) = 0. \tag{2.52}$$

However, is the converse true? If $s = i\omega_0$ is a solution of

$$F^{(1)}(s, h) = 0 \tag{2.53}$$

clearly it is also a solution of

$$F^{(1)}(-s, h) = 0 \tag{2.54}$$

but are $F(s, h)$ and $F(-s, h)$ also zero? Inverting (2.51) and (2.52):

$$\{A_0(s)A_0(-s) - A_n(s)A_n(-s)\}F(s, h) = A_0(s)F^{(1)}(s, h) + A_n(s)z^n F^{(1)}(-s, h)$$

$$\{A_0(s)A_0(-s) - A_n(s)A_n(-s)\}F(-s, h) = A_0(-s)F^{(1)}(-s, h) + A_n(-s)z^{-n} F^{(1)}(s, h).$$

Thus $F^{(1)}(s, h) = 0$ implies that *either* $F(s, h) = F(-s, h) = 0$ *or* $A_0^{(1)}(s) \equiv A_0(s)A_0(-s) - A_n(s)A_n(-s) = 0$.

Generalizing this result we have that $F^{(r+1)}(s, h) = 0$ implies that either $F^{(r)}(s, h) = 0$ or $A_0^{(r+1)} = 0$. Consequently we may deduce that if $W(\omega_0^2) = 0$ then $s = i\omega_0$ is a solution of $F(s, h) = 0$ provided that

$$\alpha(s) \equiv A_0^{(1)}(s)A_0^{(2)}(s) \ldots A_0^{(n+1)}(s) \neq 0 \tag{2.55}$$

at $s = i\omega_0$. If $\alpha(i\omega_0) = 0$ then $s = i\omega_0$ may or may not be a solution and it will be necessary to test this directly.

It remains to obtain the analogue of the $W'(\omega_0^2)$ criterion for the single-delay case to determine whether a crossing point is stabilizing or destabilizing. To derive this, we first show that the direction of crossing by a root of $F(s, h) = 0$ is the same as that by the corresponding root of $F^{(1)}(s, h) = 0$ if and only if

$$A_0^{(1)}(s) \equiv A_0(s)A_0(-s) - A_n(s)A_n(-s) > 0 \tag{2.56}$$

at the crossing point $s = i\omega_0$.

Let us suppose that $s = i\omega_0$, $h = h_0$ satisfy both $F(s, h) = 0$ and $F^{(1)}(s, h) = 0$. We are interested in the behaviour of the respective roots for values of h close to h_0. So if h is perturbed slightly so that $h = h_0 + \delta h$, let δs_1 and δs_2 denote the perturbations of the two roots so that

$$\left. \begin{array}{l} F(i\omega_0 + \delta s_1, h_0 + \delta h) = 0 \\ F^{(1)}(i\omega_0 + \delta s_2, h_0 + \delta h) = 0. \end{array} \right\} \tag{2.57}$$

Expanding these, and retaining only the lowest-order terms, we obtain

$$\left. \begin{array}{l} \delta s_1 F_s(i\omega_0, h_0) + \delta h F_h(i\omega_0, h_0) = 0 \\ \delta s_2 F_s^{(1)}(i\omega_0, h_0) + \delta h F_h^{(1)}(i\omega_0, h_0) = 0 \end{array} \right\} \tag{2.58}$$

in which the subscripts s and h denote partial differentiation with respect to the corresponding argument of $F(s, h)$.

Differentiating equation (2.44) partially with respect to s and h, and putting $s = i\omega_0$, $h = h_0$, gives

$$\left. \begin{array}{l} F_s^{(1)}(i\omega_0, h_0) = A_0(-i\omega_0) F_s(i\omega_0, h_0) + A_n(i\omega_0) z^n F_s(-i\omega_0, h_0) \\ F_h^{(1)}(i\omega_0, h_0) = A_0(-i\omega_0) F_h(i\omega_0, h_0) - A_n(i\omega_0) z^n F_h(-i\omega_0, h_0). \end{array} \right\} \tag{2.59}$$

Thus

$$A_0(-i\omega_0) \{\delta s_2 F_s(i\omega_0, h_0) + \delta h F_h(i\omega_0, h_0)\}$$
$$+ A_n(i\omega_0) z^n \{\delta s_2 F_s(-i\omega_0, h_0) - \delta h F_h(-i\omega_0, h_0)\} = 0. \tag{2.60}$$

Then making use of (2.58a) and its complex conjugate

$$\delta s_1^* F_s(-i\omega_0, h_0) + \delta h F_h(-i\omega_0, h_0) = 0, \tag{2.61}$$

in which δs_1^* denotes the complex conjugate of δs_1, we may eliminate $F_h(\pm i\omega_0, h_0)$ from (2.60) and obtain

$$A_0(-i\omega_0) F_s(i\omega_0, h_0) \{\delta s_2 - \delta s_1\}$$
$$+ A_n(i\omega_0) z^n F_s(-i\omega_0, h_0) \{\delta s_2 + \delta s_1^*\} = 0. \tag{2.62}$$

Taking complex conjugates

$$A_0(i\omega_0) F_s(-i\omega_0, h_0) \{\delta s_2^* - \delta s_1^*\}$$
$$+ A_n(-i\omega_0) z^{-n} F_s(i\omega_0, h_0) \{\delta s_2^* + \delta s_1\} = 0 \tag{2.63}$$

and hence, eliminating $F_s(\pm i\omega_0, h_0)$,

$$A_0(i\omega_0) A_0(-i\omega_0) \{\delta s_2 - \delta s_1\} \{\delta s_2^* - \delta s_1^*\}$$
$$= A_n(i\omega_0) A_n(-i\omega_0) \{\delta s_2 + \delta s_1^*\} \{\delta s_2^* + \delta s_1\}. \tag{2.64}$$

Writing $\delta s_1 = x_1 + iy_1$, $\delta s_2 = x_2 + iy_2$ this may be written as

$$[A_0(i\omega_0) A_0(-i\omega_0) - A_n(i\omega_0) A_n(-i\omega_0)] (x_1^2 + x_2^2 + (y_1 - y_2)^2)$$
$$= 2x_1 x_2 [A_0(i\omega_0) A_0(-i\omega_0) + A_n(i\omega_0) A_n(-i\omega_0)]. \tag{2.65}$$

As the second expression on each side of this equation is positive it follows that x_1 and x_2 are of the same sign if and only if

$$A_0^{(1)}(i\omega_0) \equiv A_0(i\omega_0)A_0(-i\omega_0) - A_n(i\omega_0)A_n(-i\omega_0) > 0 \tag{2.66}$$

and are of opposite sign if and only if

$$A_0^{(1)}(i\omega_0) < 0. \tag{2.67}$$

Thus the direction of crossing by a root of $F(s, h) = 0$ is the same as that by the corresponding root of $F^{(1)}(s, h) = 0$ if and only if $A_0^{(1)}(s) > 0$ at the crossing point $s = i\omega_0$ and is opposite if and only if $A_0^{(1)}(s) < 0$.

Thus for multiple commensurate delays we have that

$$S \equiv \operatorname{sgn} \operatorname{Re} \left. \frac{ds}{dh} \right|_{s=i\omega_0} = \operatorname{sgn}\{\alpha(i\omega_0)W'(\omega_0^2)\} \tag{2.68}$$

where, as before,

$$\alpha(s) = A_0^{(1)}(s)A_0^{(2)}(s) \ldots A_0^{(n-1)}(s). \tag{2.69}$$

Example 2.10 $F(s, h) = s + 1 + \exp(-sh) + 3\exp(-2sh)$

(a) $F(s, 0) = s + 5$, hence stable at $h = 0$.
(b) Proceeding as above:

$$F(s, h) = s + 1 + z + 3z^2 = 0$$

$$F(-s, h) = -s + 1 + z^{-1} + 3z^{-2} = 0.$$

Eliminating z^2 using (2.44) gives

$$F^{(1)}(s, h) = (-s^2 - 8) + (-s - 2)z$$

and so,

$$A_0^{(1)}(s) = (-s^2 - 8) \text{ and } A_1^{(1)}(s) = (-s - 2)$$

and hence

$$A_0^{(1)}(s)A_0^{(1)}(-s) - A_1^{(1)}(s)A_1^{(1)}(-s) = s^4 + 17s^2 + 60 = (s^2 + 5)(s^2 + 12)$$

and with $s = i\omega$,

$$W(\omega^2) = (5 - \omega^2)(12 - \omega^2) = 0.$$

Hence $\omega_1^2 = 5$, $\omega_2^2 = 12$ and

$$W'(\omega^2) = 2\omega^2 - 17 = -7 \text{ at } \omega_1, \text{ and } +7 \text{ at } \omega_2.$$

Now

$$\frac{1}{z} = \frac{2 + i\omega}{\omega^2 - 8} = \exp(i\omega h)$$

$$\therefore \text{ at } \omega_1 = \sqrt{5}, \quad \omega_1 h = \pi + \tan^{-1}\frac{\sqrt{5}}{2} + 2q\pi$$

$$\therefore h_1(q) = \frac{\tan^{-1}\frac{\sqrt{5}}{2} + (2q+1)\pi}{\sqrt{5}}.$$

The sign of $W'(5) < 0$, and $A_0^{(1)}(i\sqrt{5}) < 0$, hence $S > 0$, so that these values of h are destabilizing. At

$$\omega_2 = \sqrt{12} = 2\sqrt{3}, \quad \exp(i\omega_2 h) = \frac{2 + i2\sqrt{3}}{4} = \exp(i\pi/3)$$

hence

$$h_2(q) = \frac{\left(2q + \frac{1}{3}\right)\pi}{2\sqrt{3}}.$$

The sign of $W'(12)$ is positive, as is the sign of $A_0^{(1)}(i2\sqrt{3})$. Hence, $h_2(q)$ are also destabilizing.

The stability interval is $0 \leq h < \min(h_2(0), h_1(0))$. Now $h_2(0) = \pi/6\sqrt{3} = 0.3023$; $h_1(0) = 1.7811$. Hence this commensurate delay system is stable for $0 \leq h < \pi/6\sqrt{3}$ only.

The methods of this section are directly applicable to the matrix case with characteristic equations: $\det[sI - \sum A_m \exp(-smh)] = 0$.

2.8 COMMENSURATE DELAYS AND THE METHOD OF STEPS

At the start of this chapter it was shown that the method of steps was a systematic technique for finding the response, or an error signal, for a time-delay system with a single delay. With little modification the technique applies also to commensurate systems.

Consider, for example, the system with error transfer function

$$\frac{1}{1 + \alpha(s)z + \beta(s)z^3},$$

where α and β are delay-free transfer functions.

Clearly this may be expanded formally as a power series in powers of z, with a corresponding summation of delayed time functions in the time domain.

If delays are not related by integers it is still possible to use the method of steps but the step lengths would not be equal (not having a common denominator) and it is unlikely, if not impossible, that such expansion could be expressed simply.

3

The method using a generalized Heaviside expansion

3.1 INTRODUCTION

In section 1.3, constant parameter linear delay-free systems are considered and in particular it is shown that the error $e(t)$ can be written as a sum of exponential terms (see equation (1.5)). This is referred to as the Heaviside expansion formula for $e(t)$. The coefficients occurring in the exponents are the roots of the characteristic equation which either are real or occur in complex conjugate pairs. Moreover, in order for the system to be asymptotically stable, each root must have a negative real part.

The latter part of Chapter 1 is devoted to the evaluation of cost functionals for such systems and in particular that of the integral of square error

$$J = \int_0^\infty e^2(t) \, dt. \tag{3.1}$$

Of the several methods presented, the one of particular interest for the purposes of this chapter is that due to R.S. Phillips for delay-free systems and is to be found in detail in the book by James, Nichols and Phillips (1947). Brief details of the method are given in Chapter 1. In particular, it is shown how the integral can be written as a double summation involving the roots p_i of the characteristic equation. Then, using the properties of the roots of polynomial equations, it is possible to evaluate these summations without explicit knowledge of the roots.

In this chapter it will be shown that extensions of these ideas may also be used for systems *with* time delay. As shown in Chapter 2, the major difference is that the characteristic equation for such systems possesses an infinite number of roots.

Consequently the summations that arise will no longer have a finite number of terms.

The basic approach of this chapter was first presented in 1982 at the Third IFAC–AFCET Symposium on Control of Distributed Parameter Systems (Gorecki and Popek, 1983). Until that time, it was in fact believed in certain circles that it was *not* possible to evaluate cost functionals for time-delay systems in closed-form. Strictly speaking, this was not the first time such evaluations had been carried out since Repin (1965) had evaluated integrals analogous to those arising in simple systems with a single time delay. However, his result was somewhat obscured and its full implications were not appreciated until much later, as will be shown in Chapter 5.

The first step in the method is to obtain the generalized Heaviside expansion for the error $e(t)$. As for the delay-free case, this is obtained from the general formula for the inverse Laplace transform using the theory of residues (see (1.5)). Substitution of this expansion into the cost functional (3.1) followed by integration with respect to time results in an expression for J involving two infinite summations (see (1.18)). Exactly as in the delay-free case, a necessary condition for J to exist is that the exponential terms in the Heaviside expansion decay with time. This is equivalent to the condition that all the roots of the characteristic equation lie in the left half of the complex plane.

The final step in the evaluation procedure is the calculation of the infinite summations. Again, exactly as in the delay-free case, it will be shown that it is possible to evaluate these summations without the explicit knowledge of the characteristic roots p_i and moreover that the resulting expressions involve only coefficients arising in the expression for the error transform $E(s)$.

In order to demonstrate the key steps in the evaluation procedure, a simple example will be considered in detail. This will be followed by the general case of a system containing a single time delay, as exemplified in (2.12). The remaining section of this chapter will then consider the extension of the method to the evaluation of other integral criteria.

3.2 THE SINGLE-DELAY PROBLEM: A SIMPLE EXAMPLE

For demonstration purposes, we consider the system whose error e(t) has Laplace transform of the form

$$E(s) = \frac{1}{s + Ke^{-sh}}. \tag{3.2}$$

This is the special case of (2.12) with $A(s) = s$, $B(s) = 1$, $C(s) = K$ and $D(s) = 0$. The starting point for the method is to obtain the generalized Heaviside expansion for $e(t)$, which, using the formula for the inverse Laplace transform, is given by

$$e(t) = \frac{1}{2\pi i} \int_{-i\infty}^{i\infty} E(s)e^{st} \, ds = \frac{1}{2\pi i} \int_{-i\infty}^{i\infty} \frac{e^{st}}{s + Ke^{-sh}} \, ds. \tag{3.3}$$

This latter integral may be evaluated by using contour integration and the theory of residues (Churchill, Brown and Verhay, 1974). Firstly the contour is closed in the left half-plane (noting, in passing, that the contribution from the infinite semi-circle is indeed zero). Thus

$$e(t) = \sum_i \text{res} \left(\frac{e^{st}}{s + Ke^{-sh}} \right) \tag{3.4}$$

in which the summation is taken over all the poles of $E(s)$, or equivalently all the solutions $s = s_i$ of the characteristic equation

$$s + Ke^{-sh} = 0. \tag{3.5}$$

Recall the stability example 2.1, that if $0 \leq hK < \pi/2$, which we assume satisfied, then all of these poles lie in the left half-plane.

We now make use of the standard result that if $f(s) = p(s)/q(s)$ and s_0 is a simple zero of q where $p(s_0) \neq 0$ then the residue of f at s_0 is $p(s_0)/q'(s_0)$. Then, using the fact that

$$\frac{d}{ds}(s + Ke^{-sh}) = 1 - Khe^{-sh} \tag{3.6}$$

(3.4) may be written

$$e(t) = \sum_i \frac{e^{s_i t}}{(1 - Khe^{-s_i h})} H(t) \tag{3.7a}$$

$$= \sum_i \frac{e^{s_i t}}{(1 + hs_i)} H(t) \tag{3.7b}$$

since, at a characteristic root, (3.5) is satisfied and so $Ke^{-s_i h} = -s_i$. (3.7b) may be compared with (1.5), the Heaviside expansion for the impulse response in the delay-free case, and it may be seen that they are of similar form but with A_i being replaced by $(1 + hs_i)^{-1}$ and the summation now having an infinite number of terms.

To obtain an expression for the cost functional J, the above formula for $e(t)$ is substituted into (3.1) resulting in

$$J = \int_0^\infty \left(\sum_i \frac{e^{s_i t}}{1 + hs_i} \right) \left(\sum_j \frac{e^{s_j t}}{1 + hs_j} \right) dt$$

$$= \sum_i \frac{1}{1 + hs_i} \left\{ \sum_j \frac{1}{1 + hs_j} \int_0^\infty e^{(s_i + s_j)t} dt \right\}. \tag{3.8}$$

Evaluation of the inner integral then yields

$$J = -\sum_i \frac{1}{1 + hs_i} \left\{ \sum_j \frac{1}{(1 + hs_j)(s_i + s_j)} \right\} \tag{3.9}$$

since $\text{Re}(s_i + s_j) < 0$. This is the analogue of the delay-free equation (1.18).

Alternatively, with the use of partial fractions, this expression may be written

$$J = -\sum_i \frac{1}{(1+hs_i)(1-hs_i)} \left\{ \sum_j \left(\frac{1}{s_j + s_i} - \frac{1}{s_j + h^{-1}} \right) \right\}. \tag{3.10}$$

The next step is to evaluate the inner sum and it may be seen that both parts are of the form

$$\sum (s_j + \alpha)^{-1}$$

for some constant α. Consequently it suffices to evaluate this general summation.

This goal may be achieved by using the Weierstrass decomposition for an entire function $F(s)$ of class A (Levin, 1964), namely

$$F(s) = e^{\chi s} \prod_j \left(1 - \frac{s}{s_j}\right) \tag{3.11}$$

where χ is a real number and $s = s_j$ are the zeros of $F(s)$. It will be seen shortly that the value of the number χ is immaterial since it does not appear in the final result (Gorecki and Popek, 1984). Logarithmic differentiation of (3.11) then gives

$$\frac{F'(s)}{F(s)} = \chi - \sum_j \frac{1}{s_j - s}. \tag{3.12}$$

The usefulness of the Weierstrass decomposition can now be clearly seen since the summation occurring in (3.12) is of precisely the required form for the evaluation of (3.10).

Since the summation in (3.12) is taken over the zeros of $F(s)$, the appropriate choice for this function is

$$F(s) = s + Ke^{-sh} \tag{3.13}$$

and moreover this satisfies the relevant conditions. With this, (3.12) becomes

$$\sum_j \frac{1}{s_j - s} = \chi - \frac{1 - hKe^{-sh}}{s + Ke^{-sh}} \tag{3.14}$$

and consequently, putting $s = -s_i$ and using $Ke^{-s_i h} = -s_i$, we obtain

$$\sum_j \frac{1}{s_j + s_i} = \chi + \frac{s_i + hK^2}{s_i^2 + K^2}. \tag{3.15}$$

Secondly, $s = -h^{-1}$ implies

$$\sum_j \frac{1}{s_j + h^{-1}} = \chi + h. \tag{3.16}$$

Combining these gives

$$\sum_j \left(\frac{1}{s_j + s_i} - \frac{1}{s_j + h^{-1}} \right) = \frac{(1 - hs_i)s_i}{K^2 + s_i^2} \tag{3.17}$$

and so (3.10) reduces to

$$J = -\sum_i \frac{s_i}{(1 + hs_i)(s_i^2 + K^2)}. \tag{3.18}$$

At this point it is worth drawing attention to the cancelling of the terms involving the real number χ that occurred in the derivation of (3.17).

The final step in the evaluation procedure is that of evaluating the summation in (3.18) and this is achieved in an identical manner to that used in evaluating the previous summation. Two steps are required. The first is the use of partial fractions and the second is the use of the Weierstrass decomposition. Using partial fractions (3.18) becomes

$$J = \sum_i \left(\frac{1}{(1 + K^2 h^2)(s_i + h^{-1})} - \frac{1}{2(1 - ihK)(s_i + iK)} - \frac{1}{2(1 + ihK)(s_i - iK)} \right). \tag{3.19}$$

These summations are exactly of the same form considered earlier and consequently

$$\sum_i \frac{1}{s_i + h^{-1}} = \chi + h. \tag{3.20}$$

Also substituting $s = -iK$ and $s = iK$ into (3.14) in turn gives

$$\sum_i \frac{1}{s_i + iK} = \chi - \frac{1 - hKe^{ihK}}{-iK + Ke^{ihK}},$$

$$\sum_i \frac{1}{s_i - iK} = \chi - \frac{1 - hKe^{-ihK}}{iK + Ke^{-ihK}}. \tag{3.21}$$

The final expression for J is therefore

$$J = \frac{\chi + h}{1 + K^2 h^2} - \frac{1}{2(1 - ihK)} \left(\chi - \frac{1 - hKe^{ihK}}{-iK + Ke^{ihK}} \right)$$

$$- \frac{1}{2(1 + ihK)} \left(\chi - \frac{1 - hKe^{-ihK}}{-iK + Ke^{ihK}} \right) = \frac{\cos hK}{2K(1 - \sin hK)} \tag{3.22}$$

after some algebraic simplification. Again it is noted that although the number χ occurs in each of the summations, these contributions all cancel and χ does not occur in the final expression.

3.3 THE SINGLE-DELAY PROBLEM: THE GENERAL CASE

As stated in Chapter 2, the standard form for error in the single-delay problem is that having Laplace transform (2.12)

$$E(s) = \frac{B(s) + D(s)e^{-sh}}{A(s) + C(s)e^{-sh}} \tag{3.23}$$

in which $A(s)$, $B(s)$, $C(s)$ and $D(s)$ are polynomials of finite degree and with real coefficients. Within the present context, asymptotic stability is a necessary condition for the ISE to exist, which is one reason why the subject was considered in detail in Chapter 2. However, it is not sufficient. A further condition is required and this related to the degrees of the polynomials $A(s)$ to $D(s)$. One such choice is that the degree of $A(s)$ be greater than the degrees of $B(s)$, $C(s)$ and $D(s)$ and as this is usually satisfied in practical situations this is the choice that will be made. Thus we henceforth assume that the system is asymptotically stable, which can be checked by the direct method of Chapter 2, and that the polynomial $A(s)$ dominates.

It will now be shown that the procedure used above for the evaluation of the cost functional J for the particular simple example applies equally well for that of the more general system, as described by (3.23). However, before demonstrating this, it is worth summarizing the basic steps involved in the procedure. These are as follows:

(i) Express $e(t)$ as a generalized Heaviside expansion (see (3.7)).
(ii) Perform the integration over t to enable J to be written as a double infinite summation (see (3.9)).
(iii) Use partial fractions to write the inner summation as one involving only single terms (see (3.10)).
(iv) Use the Weierstrass decomposition to obtain algebraic expressions for these inner summations and hence obtain J in the form of a single infinite summation (see (3.18)).
(v) Repeat steps (iii) and (iv) this time for evaluation of the remaining summation.

For the general case $E(s)$, the Laplace transform of the error $e(t)$, is given by (3.23) and so by analogy with (3.7a), $e(t)$ is given by the generalized Heaviside expansion

$$e(t) = \sum_i \left\{ \frac{B(s_i) + D(s_i)e^{-s_ih}}{A'(s_i) + C'(s_i)e^{-s_ih} - hC(s_i)e^{-s_ih}} \right\} e^{s_it}. \qquad (3.24)$$

The summation is taken over all the poles of $E(s)$ or equivalently over all the solutions $s = s_i$ of the characteristic equation

$$A(s) + C(s)e^{-sh} = 0. \qquad (3.25)$$

Since the system is assumed to be stable, all of these lie in the left half-plane. This can be checked in particular examples by the direct method of Chapter 2. It follows that

$$e^{-s_ih} = -\frac{A(s_i)}{C(s_i)}, \quad C(s_i) \neq 0, \qquad (3.26)$$

since we also assume that $A(s)$ and $C(s)$ have no common factors. Hence (3.24) may be written

$$e(t) = \sum_i \frac{L(s_i)e^{s_it}}{M(s_i)} \qquad (3.27)$$

in which $L(s)$ and $M(s)$ are polynomials in s given by

$$\left.\begin{array}{l}L(s) = B(s)C(s) - A(s)D(s),\\ M(s) = A'(s)C(s) - A(s)C'(s) + hA(s)C(s).\end{array}\right\} \quad (3.28)$$

The cost functional J is then given by

$$J = -\sum_i \frac{L(s_i)}{M(s_i)} \left\{ \sum_j \frac{L(s_j)}{M(s_j)(s_j + s_i)} \right\}. \quad (3.29)$$

This is the generalization of (3.9) and so concludes step (ii).

For step (iii) what is required is a partial fraction expansion for the expression under the inner summation sign in (3.29). However, since in what follows, further expressions of this form will be encountered, the more general expression $L(s)/(M(s)(s-a))$ is considered. In this the parameter a is arbitrary. The corresponding partial fraction expansion is

$$\frac{L(s)}{M(s)(s-a)} = \frac{L(a)}{M(a)(s-a)} + \sum_k \frac{L(p_k)}{M'(p_k)(p_k - a)(s - p_k)} \quad (3.30)$$

in which $s = p_k$ denotes the zeros of $M(s)$. Taking $s = s_j$ and summing then yields

$$\sum_j \frac{L(s_j)}{M(s_j)(s_j - a)} = \frac{L(a)}{M(a)} \sum_j \frac{1}{(s_j - a)}$$

$$+ \sum_k \frac{L(p_k)}{M'(p_k)(p_k - a)} \left\{ \sum_j \frac{1}{(s_j - p_k)} \right\}. \quad (3.31)$$

Thus, exactly as in the simple example considered earlier, it is necessary to evaluate summations of the form

$$\sum_j (s_j - a)^{-1}.$$

As before, expressions for these may be obtained with the aid of the Weierstrass decomposition and in particular (3.12) which may be written

$$\sum_j \frac{1}{s_j - a} = \chi - \frac{F'(a)}{F(a)} \quad (3.32)$$

in which the s_j are all the zeros of the function $F(s)$ and consequently $F(s)$ is taken to be

$$F(s) = A(s) + C(s)e^{-sh}. \quad (3.33)$$

Thus (3.32) becomes

$$\sum_j \frac{1}{s_j - a} = \chi - \frac{A'(a) + C'(a)e^{-ah} - hC(a)e^{-ah}}{A(a) + C(a)e^{-ah}} \quad (3.34)$$

and hence, also putting $a = p_k$ in (3.34), the summation in (3.31) may be written in the form

$$\sum_j \frac{L(s_j)}{M(s_j)(s_j - a)} = \chi \left\{ \frac{L(a)}{M(a)} + \sum_k \frac{L(p_k)}{M'(p_k)(p_k - a)} \right\}$$

$$- \frac{L(a)}{M(a)} \left\{ \frac{A'(a) + C'(a)e^{-ah} - hC(a)e^{-ah}}{A(a) + C(a)e^{-ah}} \right\}$$

$$- \sum_k \frac{L(p_k)}{M'(p_k)(p_k - a)} \left\{ \frac{A'(p_k) + C'(p_k)e^{-p_k h} - hC(p_k)e^{-p_k h}}{A(p_k) + C(p_k)e^{-p_k h}} \right\} \quad (3.35)$$

determined by the function $F(s)$ which is the same in both cases.

In the simple example considered the term in χ was absent. This is also true here. We show that the coefficient of χ in (3.35) is identically zero. To do this, we first note that, from (3.30),

$$\frac{L(s)}{M(s)} = \frac{L(a)}{M(a)} + \sum_k \frac{L(p_k)(s - a)}{M'(p_k)(p_k - a)(s - p_k)} \quad (3.36)$$

and hence

$$\frac{L(a)}{M(a)} + \sum_k \frac{L(p_k)}{M'(p_k)(p_k - a)} = \lim_{s \to \infty} \left\{ \frac{L(s)}{M(s)} \right\} \quad (3.37)$$

and so from (3.28), retaining only the dominant term in $M(s)$

$$\frac{L(a)}{M(a)} + \sum_k \frac{L(p_k)}{M'(p_k)(p_k - a)} = \lim_{s \to \infty} \left\{ \frac{B(s)C(s) - A(s)D(s)}{hA(s)C(s)} \right\}$$

$$= \frac{1}{h} \lim_{s \to \infty} \left\{ \frac{B(s)}{A(s)} - \frac{D(s)}{C(s)} \right\} = 0 \quad (3.38)$$

provided that $\deg B < \deg A$ and $\deg D < \deg C$. Thus the coefficient of χ is indeed zero. These conditions will be discussed in more detail later in this section.

(3.35) may, in fact, be simplified even further by making use of the fact that $M(p_k) = 0$ and so from (3.28)

$$A'(p_k) + (C'(p_k) - hC(p_k))e^{-p_k h} = \frac{A'(p_k)}{A(p_k)} \{A(p_k) + C(p_k)e^{-p_k h}\}. \quad (3.39)$$

Thus (3.35) reduces a

$$\sum_j \frac{L(s_j)}{M(s_j)(s_j - a)} = -\frac{L(a)}{M(a)} \left\{ \frac{A'(a) + C'(a)e^{-ah} - hC(a)e^{-ah}}{A(a) + C(a)e^{-ah}} \right\}$$

$$- \sum_k \frac{L(p_k)A'(p_k)}{M'(p_k)(p_k - a)A(p_k)}. \quad (3.40)$$

The next stage is the evaluation of the final summation in (3.40). To do this, it is first noticed that this particular summation occurs in the partial fraction expansion

$$\frac{L(a)A'(a)}{M(a)A(a)} = \sum_k \frac{L(p_k)A'(p_k)}{M'(p_k)A(p_k)(a - p_k)} + \sum_i \frac{L(r_i)}{M(r_i)(a - r_i)} \quad (3.41)$$

in which the r_i are the zeros of $A(s)$ and so $A(r_i) = 0$. Moreover at such points

$$L(r_i) = B(r_i)C(r_i), \quad M(r_i) = A'(r_i)C(r_i) \quad (3.42)$$

and consequently

$$\sum_i \frac{L(r_i)}{M(r_i)(a - r_i)} = \sum_i \frac{B(r_i)}{A'(r_i)(a - r_i)} = \frac{B(a)}{A(a)} \quad (3.43)$$

since this is exactly the partial fraction expansion of $B(a)/A(a)$. Consequently (3.41) may now be rewritten in the form

$$\sum_k \frac{L(p_k)A'(p_k)}{M'(p_k)A(p_k)(p_k - a)} = \frac{B(a)}{A(a)} - \frac{L(a)A'(a)}{M(a)A(a)}$$

$$= -\frac{[C'(a) - hC(a)]B(a) - A'(a)D(a)}{M(a)} \quad (3.44)$$

making use of the expression for $L(a)$ given in (3.28). This is the required expression for the final summation in (3.40) and consequently this now becomes

$$\sum_j \frac{L(s_j)}{M(s_j)(s_j - a)} = -\frac{L(a)}{M(a)} \left\{ \frac{A'(a) + C'(a)\mathrm{e}^{-ah} - hC(a)\mathrm{e}^{-ah}}{A(a) + C(a)\mathrm{e}^{-ah}} \right\}$$

$$+ \frac{[C'(a) - hC(a)]B(a) - A'(a)D(a)}{M(a)} \quad (3.45)$$

which, after some algebraic simplification again making use of (3.28) for $L(a)$ and $M(a)$, finally reduces to

$$\sum_j \frac{L(s_j)}{M(s_j)(s_j - a)} = -\frac{B(a) + D(a)\mathrm{e}^{-ah}}{A(a) + C(a)\mathrm{e}^{-ah}}. \quad (3.46)$$

This particular equation is extremely useful and will be used again later. However, for present purposes it is the particular form of this when $a = -s_i$ that is required in (3.29), the equation for the cost functional J. However, when $a = -s_i$, from (3.26)

$$\mathrm{e}^{s_i h} = -\frac{C(s_i)}{A(s_i)}, \quad A(s_i) \neq 0 \quad (3.47)$$

and so

$$\sum_j \frac{L(s_j)}{M(s_j)(s_j + s_i)} = -\frac{A(s_i)B(-s_i) - C(s_i)D(-s_i)}{A(s_i)A(-s_i) - C(s_i)C(-s_i)} \quad (3.48)$$

and hence J may be written

$$J = \sum_i \frac{L(s_i)G(s_i)}{M(s_i)F(s_i)} \quad (3.49)$$

in which

$$F(s) = A(s)A(-s) - C(s)C(-s), \quad G(s) = A(s)B(-s) - C(s)D(-s). \qquad (3.50)$$

This concludes step (iv) since J is now in the form of a single infinite summation and (3.49) may be compared with the corresponding expression (3.18) obtained when considering the simple example. The only extra complication arising when considering the general case is the evaluation of the final summation in (3.40) and as was shown above this can be achieved without too much difficulty. It remains finally to do step (v), the evaluation of the summation in (3.49) and, in principle, this is achieved in the same manner as the previous summation. In fact much use of this may be made if use is made of the partial fraction expansion of $G(s_i)/F(s_i)$, which is

$$\frac{G(s_i)}{F(s_i)} = \sum_m \frac{G(u_m)}{F'(u_m)} \frac{1}{s_i - u_m} \qquad (3.51)$$

in which u_m denotes the zeros of $F(s)$, which since $F(s)$ is the finite polynomial given in (3.50) will be finite in number. Inserting (3.51) into (3.49) and interchanging the order of summation leads to

$$J = \sum_m \frac{G(u_m)}{F'(u_m)} \left\{ \sum_i \frac{L(s_i)}{M(s_i)(s_i - u_m)} \right\}. \qquad (3.52)$$

The inner summation in this equation is of exactly the same form considered previously and an expression for which is given in (3.46). Substituting this into (3.52) then gives

$$J = -\sum_m \frac{G(u_m)}{F'(u_m)} \left\{ \frac{B(u_m) + D(u_m)e^{-u_m h}}{A(u_m) + C(u_m)e^{-u_m h}} \right\}. \qquad (3.53)$$

This is the final expression for J and so concludes step (v). It is seen that the resulting summation is a finite one, being taken over the zeros of $F(s)$, which are given by

$$F(u_m) \equiv A(u_m)A(-u_m) - C(u_m)C(-u_m) = 0. \qquad (3.54)$$

In order to perform the evaluation it was necessary to assume that $\deg B < \deg A$ and $\deg D < \deg C$. As stated at the beginning of this section we consider only those systems in which $A(s)$ dominates and so the first condition is automatically satisfied. The second condition is then equivalent to the condition that $\deg L < \deg M$. Such a condition is necessary for the summations given in (3.29) to exist. If it were not satisfied then the derivation of the form of J given in (3.29) would not be valid. It is also worth noting that if $\deg L > \deg M$ then many of the partial fraction expansions would also be invalid.

3.4 EXTENSION TO OTHER INTEGRAL CRITERIA

Several other integral criteria were introduced in Chapter 1 and it was shown there how these could be evaluated in the delay-free case. In principle it is possible to evaluate such integrals for systems involving time-delay by using extensions of the techniques presented in this chapter. However, we demonstrate these only for the example of weighted quadratic criteria starting with the integral

$$J_r = \int_0^\infty t^r e^2(t)\,dt \qquad (3.55)$$

for any positive integer r. Although it is, in principle, possible to do this for any such integer r, for simplicity only the case $r = 1$ will be considered in detail. In this case, using the expression for $e(t)$ given in (3.27), it follows that

$$J_1 = \sum_i \frac{L(s_i)}{M(s_i)} \sum_j \frac{L(s_j)}{M(s_j)} \int_0^\infty t e^{(s_i+s_j)t}\,dt$$

$$= \sum_i \frac{L(s_i)}{M(s_i)} \sum_j \frac{L(s_j)}{M(s_j)(s_i+s_j)^2}. \qquad (3.56)$$

Thus in contrast with the $r = 0$ case, the inner sum now involves $(s_i + s_j)^2$. However, since (3.46) is valid for any a it follows that

$$\sum_j \frac{L(s_j)}{M(s_j)(s_j-a)^2} = \frac{\partial}{\partial a} \sum_j \frac{L(s_j)}{M(s_j)(s_j-a)}$$

$$= -\frac{\partial}{\partial a}\left\{\frac{B(a)+D(a)e^{-ah}}{A(a)+C(a)e^{-ah}}\right\}$$

$$= -\frac{B'(a)+D'(a)e^{-ah}-hD(a)e^{-ah}}{A(a)+C(a)e^{-ah}}$$

$$+ \frac{(B(a)+D(a)e^{-ah})(A'(a)+C'(a)e^{-ah}-hC(a)e^{-ah})}{(A(a)+C(a)e^{-ah})^2} \qquad (3.57)$$

and hence putting $a = -s_i$ and making use of the fact that $e^{s_i h} = -C(s_i)/A(s_i)$,

$$\sum_j \frac{L(s_j)}{M(s_j)(s_j+s_i)^2} = -\frac{P(s_i)}{F(s_i)} + \frac{Q(s_i)}{[F(s_i)]^2} \qquad (3.58)$$

where $F(s)$ is given by (3.50) and

$$P(s) = A(s)B'(-s) - C(s)D'(-s) + hC(s)D(-s)$$

$$Q(s) = [A(s)B(-s) - C(s)D(-s)][A(s)A'(-s) - C(s)C'(-s)]$$

$$+ hC(s)C(-s)]. \qquad (3.59)$$

Substituting (3.58) into (3.56) then gives

$$J_1 = \sum_i \frac{L(s_i)}{M(s_i)}\left\{-\frac{P(s_i)}{F(s_i)} + \frac{Q(s_i)}{[F(s_1)]^2}\right\}. \qquad (3.60)$$

Now in analogy with (3.51)

$$\frac{P(s_i)}{F(s_i)} = \sum_m \frac{P(u_m)}{F'(u_m)} \frac{1}{(s_i - u_m)}$$

$$\frac{Q(s_i)}{[F(s_i)]^2} = \tag{3.61}$$

$$= \sum_m \left\{ \frac{Q(u_m)}{[F'(u_m)]^2} \frac{1}{(s_i - u_m)^2} + \frac{Q'(u_m)}{[F'(u_m)]^2} \frac{1}{(s_i - u_m)} - \frac{F''(u_m)Q(u_m)}{[F'(u_m)]^3} \frac{1}{(s_i - u_m)} \right\}$$

and hence, interchanging the order of summation

$$J_1 = \sum_m \left\{ \left[-\frac{P(u_m)}{F'(u_m)} + \frac{Q'(u_m)}{[F'(u_m)]^2} - \frac{F''(u_m)Q(u_m)}{[F'(u_m)]^3} \right] \sum_i \frac{L(s_i)}{M(s_i)} \frac{1}{(s_i - u_m)} \right.$$

$$\left. + \frac{Q(u_m)}{[F'(u_m)]^2} \frac{\partial}{\partial u_m} \sum_i \frac{L(s_i)}{M(s_i)} \frac{1}{(s_i - u_m)} \right\}. \tag{3.62}$$

Consequently, again making use of the expression given in (3.46)

$$J_1 = \sum_m \left\{ \left[-\frac{P(u_m)}{F'(u_m)} + \frac{Q'(u_m)}{[F'(u_m)]^2} - \frac{F''(u_m)Q(u_m)}{[F'(u_m)]^3} \right] \left[\frac{B(u_m) + D(u_m)e^{-u_m h}}{A(u_m) + C(u_m)e^{-u_m h}} \right] \right.$$

$$\left. + \frac{Q(u_m)}{[F'(u_m)]^2} \frac{\partial}{\partial u_m} \left[\frac{B(u_m) + D(u_m)e^{-u_m h}}{A(u_m) + C(u_m)e^{-u_m h}} \right] \right\}. \tag{3.63}$$

This is the final result and may be compared with (3.53), the result in the case $r = 0$.

To demonstrate these results, the earlier example is again considered. In this $E(s) = (s + Ke^{-sh})^{-1}$ and so

$$A(s) = s, \quad B(s) = 1, \quad C(s) = K, \quad D(s) = 0. \tag{3.64}$$

Thus from (3.50)

$$F(s) = -(s^2 + K^2), \quad G(s) = s \tag{3.65}$$

and so there are two zeros of F(s) and these are given by $s = \pm iK$. So with $u_1 = iK$, $u_2 = -iK$,

$$J_0 = -\sum_{m=1}^{2} \frac{u_m}{-2u_m} \left\{ \frac{1}{u_m + e^{-u_m h}} \right\}$$

$$= \frac{1}{2} \left\{ \frac{1}{iK + e^{-ihK}} + \frac{1}{-iK + e^{ihK}} \right\} = \frac{\cos hK}{2K(1 - \sin hK)}, \tag{3.66}$$

in agreement with (3.22).

Secondly, for $r = 1$,

$$P(s) = 0, \quad Q(s) = s(s + hK^2). \tag{3.67}$$

Thus

$$J_1 = -\sum_{m=1}^{2}\left\{\left[\frac{2u_m + hK^2}{4u_m^2} + \frac{2u_m(u_m + hK^2)}{-8u_m^3}\right]\left[\frac{1}{u_m + e^{-u_m hK}}\right]\right.$$
$$\left. + \frac{u_m(u_m + hK^2)}{4u_m^2}\frac{\partial}{\partial u_m}\left[\frac{1}{u_m + e^{-u_m hK}}\right]\right\} = \frac{1 + h^2 K^2 - hK \cos hK}{4K^2(1 - \sin hK)} \quad (3.68)$$

after some algebra.

For higher values of r, the weighted quadratic integral J_r, as defined in (3.55), may be evaluated in exactly the same manner. In particular it is found that, for the simple example considered above,

$$J_2 = \frac{(1 + h^2 K^2)^2 \cos hK + [(1 + h^2 K^2)\cos hK - hK(5 + h^2 K^2)](1 - \sin hK)}{8K^3(1 - \sin hK)^2}. \quad (3.69)$$

In addition to criteria weighted by a power of t, those weighted by an exponential are also of interest. We therefore consider

$$J(\alpha) = \int_0^\infty e^{\alpha t} e^2(t)\,dt. \quad (3.70)$$

Again the techniques introduced earlier may also be used to evaluate this integral. From the Heaviside expansion (3.27) for the error $e(t)$, we have that

$$J(\alpha) = \sum_i \frac{L(s_i)}{M(s_i)} \sum_j \frac{L(s_j)}{M(s_j)} \int_0^\infty e^{\alpha t} e^{(s_i + s_j)t}\,dt$$
$$= \sum_i \frac{L(s_i)}{M(s_i)} \sum_j \frac{L(s_j)}{M(s_j)} \frac{1}{s_i + s_j + \alpha}. \quad (3.71)$$

Since the evaluation procedure now follows in exactly the same way as before, no further details will be given.

However, it is worth noting that

$$\left.\frac{\partial}{\partial \alpha} J(\alpha)\right|_{\alpha=0} = J_1 \quad (3.72)$$

and similarly higher derivatives will lead to other J_r. Consequently if $J(\alpha)$ has been determined, all the J_r may be evaluated by differentiation without further evaluation of integrals.

4

The method based on Parseval's theorem and contour integration

4.1 INTRODUCTION

In Chapter 1 the integral of square error

$$J = \int_0^\infty e^2(t)\,dt \tag{4.1}$$

and other cost functionals were introduced. In particular, several methods were given for their evaluation for delay-free systems. The first of these is that given in James, Nichols and Phillips (1947), and in Chapter 3 it was shown how the ideas presented in this method may be extended to enable the evaluation of such integrals when time delays are present. In this chapter it will be shown how a second delay-free method, namely that of Newton, Gould and Kaiser (1957), may also be extended, to advantage.

Details of this delay-free method were given in section 1.5 and, as will be seen, the basic steps in the method are the same regardless of whether or not there is a time delay. The first such step is to use Parseval's theorem and write the cost functional J, as introduced in (4.1), in the form

$$J = \frac{1}{2\pi i} \int_{-i\infty}^{i\infty} E(s)E(-s)\,ds \tag{4.2}$$

with $E(s)$ again denoting the Laplace transform of the error $e(t)$. In writing down the integrals (4.1) and (4.2), it has been assumed that they both exist and within the present context a necessary condition for this is that the system is stable. This important concept was discussed in detail for time-delay systems in Chapter 2. In particular, methods were given for determining the stability regions of a given system, and in this chapter the strong connection between stability and cost functionals will be demonstrated. However, for present purposes we merely repeat that, for a stable system, all the poles of $E(s)$ lie in the left half of the complex s-plane. In fact this is worth emphasizing since it is the cornerstone of the method to be presented in this chapter, just as it is the cornerstone of the delay-free method of Newton, Gould and Kaiser (1957). We also remark that although this condition is necessary it is not always sufficient and further conditions are required.

The second step in their method (and it will also be the second step in the method of this chapter) is to separate the integrand of (4.2) into two parts, the first of which has all its poles in the left-half of the complex s-plane (since they are the poles of $E(s)$) and the second of which has all its poles in the right half of the complex s-plane (since they arise from $E(-s)$). The particular separation for the delay-free case is given in equation (1.33). Finally, each part of the integrand is treated separately and evaluated using contour integration, and again this will be so for the time-delay case. In both cases, much use is made of the fact that the poles of $E(s)$ and $E(-s)$ lie in opposite half-planes.

Following the format of the previous chapter, this second method will be demonstrated by considering first the simple example introduced in section 3.2. This will again be followed by the general case of a system containing a single time delay. The remaining sections of this chapter concentrate on extensions of the basic method and include systems involving commensurate delays and the evaluation of other quadratic criteria.

Historically, the basic method to be presented in this chapter first appeared in Walton and Marshall (1984a). The same method with minor adjustments was also reviewed in Walton and Gorecki (1984). Extensions of the method were given in Walton, Ireland and Marshall (1986) and in Walton and Marshall (1987b).

4.2 THE SINGLE-DELAY PROBLEM: A SIMPLE EXAMPLE

To demonstrate the method for systems involving time delay, attention is first restricted to those involving only a single time delay. For additional clarity, one simple example will be considered in detail. This will be followed by a brief treatment of an arbitrary system involving a single delay. The particular example chosen (see section 3.2) is that for which

$$E(s) = \frac{1}{s + K e^{-sh}} \qquad (4.3)$$

in which h denotes the time delay and K is a positive constant. Consequently the cost functional J is given by

$$J = \frac{1}{2\pi i} \int_{-i\infty}^{i\infty} E(s)E(-s)\,ds$$

$$= \frac{1}{2\pi i} \int_{-i\infty}^{i\infty} \frac{ds}{(s + K e^{-sh})(-s + K e^{sh})}. \tag{4.4}$$

As with the delay-free case, the system is assumed to be stable and consequently all the poles of $E(s)$, or equivalently all the zeros of $(s + K e^{-sh})$ lie in the left half of the complex s-plane. From Example 2.1 the condition for this is known to be $0 \leqslant 2Kh < \pi$.

Again the aim is to separate the integrand into two parts and then to close the contour for one part on the right and the other part on the left. One way to achieve such a separation is to consider the poles of $E(s)$, or equivalently the zeros of $(s + K e^{-sh})$. At such a point

$$K e^{-sh} = -s \tag{4.5}$$

and consequently

$$\frac{1}{(-s + K e^{sh})} = -\frac{s}{(s^2 + K^2)}. \tag{4.6}$$

The integral for the cost functional J may therefore be rewritten in the form

$$J = \frac{1}{2\pi i} \int_{-i\infty}^{i\infty} \left\{ \frac{1}{(s + K e^{-sh})} \left[\frac{1}{(-s + K e^{sh})} + \frac{s}{(s^2 + K^2)} \right] \right.$$

$$\left. - \frac{s}{(s + K e^{-sh})(s^2 + K^2)} \right\} ds \tag{4.7a}$$

$$= \frac{1}{2\pi i} \int_{-i\infty}^{i\infty} \left\{ \frac{K e^{sh}}{(-s + K e^{sh})(s^2 + K^2)} - \frac{s}{(s + K e^{-sh})(s^2 + K^2)} \right\} ds. \tag{4.7b}$$

Such an approach is always possible for systems involving a single time delay. Firstly an expression for e^{-sh} at the poles of $E(s)$, in terms of only polynomials of s, is determined as demonstrated by (4.5). An expression for $E(-s)$ at such points is then obtained by substituting for the exponential terms as demonstrated by (4.6). The required separation is then obtained by rewriting the integral for the cost functional in the manner demonstrated in (4.7a). Since the term in square brackets has been formed in such a manner that it is always zero when $(s + K e^{-sh})$ is zero, it follows that this term will cancel as is demonstrated by (4.7b). Thus the desired separation has been obtained, in the sense that the first term in the integrand of (4.7b) contains no poles of $E(s)$ whereas the second term contains no poles of $E(-s)$.

Since all the zeros of $(s + K e^{-sh})$ lie in the left half of the complex s-plane and all those of $(-s + K e^{sh})$ lie in the right half, it should now be possible to close the contour on the left for the first part of the integrand and on the right for the second part. In this way it is then possible to avoid the infinite number of poles of either $E(s)$ or $E(-s)$ that would otherwise occur. However, before doing this, consideration must be given to the zeros of $(s^2 + K^2)$, namely $s = \pm iK$. Although these are not poles of the combined integrand, they are poles of the separate parts. Consequently before separating the integral and closing the contours, the contour from $-i\infty$ to $i\infty$ is first deformed in order to avoid passing through these points and, without loss of generality, this is achieved by indenting it to the left of both $s = iK$ and $s = -iK$. The contours are then closed (see Fig. 4.1) and it is noted that the contributions from the semi-circles at infinity are indeed zero. Moreover, the first integral contains no enclosed poles and consequently is identically zero. On the other hand, the only enclosed poles of the second integral are those at $s = \pm iK$. Consequently, using the theory of residues

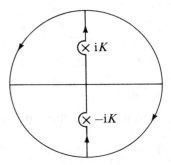

Fig. 4.1. Contours of integration.

$$J = \sum_{s=\pm iK} \mathrm{res}\left\{\frac{s}{(s + K e^{-sh})(s^2 + K^2)}\right\}$$

$$= \frac{1}{2K(i + e^{-iKh})} + \frac{1}{2K(-i + e^{iKh})} = \frac{\cos Kh}{2K(1 - \sin Kh)} \quad (4.8a)$$

in agreement with (3.22).

Alternatively (4.8a) may be written

$$J = \left\{2K \tan\left(\frac{\pi}{4} - \frac{Kh}{2}\right)\right\}^{-1} = \frac{1}{2K} \tan\left(\frac{\pi}{4} + \frac{Kh}{2}\right) \quad (4.8b)$$

from which it is easily seen that J is infinite when $Kh = \pi/2$, which we recognize as the value of Kh at which the system becomes unstable. We also note that for values of Kh greater than this the cost functional J does not exist, since the system is unstable; however, the right-hand sides of (4.8) do.

4.3 THE SINGLE-DELAY PROBLEM: THE GENERAL CASE

As shown in Chapter 2, in general, for a system containing a single time delay, $E(s)$, the Laplace transform of the error $e(t)$ will be of the form

$$E(s) = \frac{B(s) + D(s)e^{-sh}}{A(s) + C(s)e^{-sh}} \tag{4.9}$$

in which $A(s)$, $B(s)$, $C(s)$ and $D(s)$ are polynomials in s of finite degree and with real coefficients. For such a system the cost functional J is given by

$$J = \frac{1}{2\pi i} \int_{-i\infty}^{i\infty} E(s)E(-s)\,ds$$

$$= \frac{1}{2\pi i} \int_{-i\infty}^{i\infty} \left(\frac{B(s) + D(s)e^{-sh}}{A(s) + C(s)e^{-sh}}\right)\left(\frac{B(-s) + D(-s)e^{sh}}{A(-s) + C(-s)e^{sh}}\right) ds. \tag{4.10}$$

Following the above procedure, at a pole of $E(s)$,

$$e^{-sh} = -\frac{A(s)}{C(s)} \tag{4.11}$$

and consequently, at such a point

$$E(-s) \equiv \frac{B(-s) + D(-s)e^{sh}}{A(-s) + C(-s)e^{sh}} = \frac{A(s)B(-s) - C(s)D(-s)}{A(s)A(-s) - C(s)C(-s)}. \tag{4.12}$$

As before, subtracting off this term and adding it in yields

$$J = \frac{1}{2\pi i} \int_{-i\infty}^{i\infty} \left\{ E(s)\left[E(-s) - \frac{A(s)B(-s) - C(s)D(-s)}{A(s)A(-s) - C(s)C(-s)} \right] \right.$$

$$\left. + E(s)\left(\frac{A(s)B(-s) - C(s)D(-s)}{A(s)A(-s) - C(s)C(-s)}\right) \right\} ds. \tag{4.13}$$

If the contour from $-i\infty$ to $i\infty$ is now deformed to the left of *all* the zeros of $(A(s)A(-s) - C(s)C(-s))$ but remaining to the right of all the poles of $E(s)$, there will be no change in the value of the integral. As before, the contour for the first part of the integrand is now closed on the left and that for the second part on the right. Since there are no enclosed poles of the first part and the only poles of the second part occur at the solutions of

$$A(s)A(-s) - C(s)C(-s) = 0 \tag{4.14}$$

it follows that

$$J = -\sum_{r} \operatorname*{res}_{s=s_r} \left(\frac{B(s) + D(s)e^{-sh}}{A(s) + C(s)e^{-sh}}\right)\left(\frac{A(s)B(-s) - C(s)D(-s)}{A(s)A(-s) - C(s)C(-s)}\right) \tag{4.15}$$

in which $s = s_r$ denote the solutions of (4.14). Thus the problem has been reduced to finding the zeros of the finite polynomial given in (4.14) and evaluating residues at these points.

We note that this polynomial has already been encountered. This was in Chapter 2 in the context of stability where it was shown that the pure imaginary solutions of (4.14) are the potential crossing points for solutions of the characteristic equation. However, in the present chapter *all* solutions of (4.14) are of interest, whether on the imaginary axis or not, since there are contributions to the value of J from all of them.

In deriving (4.15), certain assumptions are made. The first is that there are no contributions from the semi-circular contours at infinity which arise when closing the contours. A sufficient condition for this is that the degree of the polynomial $A(s)$ be strictly greater than that of all the other three polynomials. Moreover this will usually be so in systems of practical interest. Secondly, it is assumed that it is possible to deform the contour from $-i\infty$ to $i\infty$ to the left of all the solutions of (4.13) but to remain to the right of all the poles of $E(s)$. There arises one situation when this is not possible and this occurs when both $A(s)$ and $C(s)$ possess a common root, $s = s_p$ say, in the left half-plane. Thus $s = s_p$ is a pole of $E(s)$ assuming it is not also a zero of $B(s) + D(s)\,e^{-sh}$. Moreover, it is also a solution of (4.14) and consequently it is not possible to deform the contour to the left of this point without changing the value of the integral. If such points are present, the above technique requires modification in the sense that the contour must remain to the right of these. The contribution to J from such a point must then be determined from the first part of the integrand in (4.13) and is found to be

$$J_p = \operatorname*{res}_{s=s_p} \left(\frac{D(s) + B(s)e^{sh}}{A(-s) + C(-s)e^{sh}} \right) \left(\frac{A(-s)D(-s) - B(-s)C(-s)}{A(s)A(-s) - C(s)C(-s)} \right) \quad (4.16)$$

and not the form given in (4.15). Naturally, the contributions from all the other solutions of (4.14) are obtained using (4.15). Later in this section alternative, but equivalent, forms of (4.15) will be discussed and we remark here that one of these (that given in (4.25)) is in fact more appropriate for the case just discussed.

One other case also merits further comment, and this occurs when both $A(s)$ and $C(s)$ possess a common factor s. For the cost functional J to exist, $E(s)$ must be well defined at the origin and consequently it follows that

$$B(0) + D(0) = 0 \quad (4.17)$$

Moreover, $s = 0$ will also be a double root of (4.14). However, $s = 0$ is only a single pole and not a double pole as might first be expected. The contribution to J, denoted by J_0, from this pole may be found from (4.15) by first writing

$$A(s) = s\hat{A}(s), \quad C(s) = s\hat{C}(s) \quad (4.18)$$

in which $\hat{A}(0)$ and $\hat{C}(0)$ are assumed to be not simultaneously zero. It then follows that

$$J_0 = \frac{B(0)\{B'(0) + D'(0) + hB(0)\}}{[\hat{A}^2(0) - \hat{C}^2(0)]} \quad (4.19)$$

where the prime denotes a derivative.

It is also worth noting that (4.14) is symmetric in s and consequently if $s = s_r$ is a solution then so is $s = -s_r$. Similarly since the coefficients of $A(s)$ and $C(s)$ are real, if $s = s_r$ is a solution then so is $s = s_r^*$, the complex conjugate of s_r. In particular, in the example considered above, the two solutions were $s = \pm iK$ and so both the above were satisfied. It is in fact possible to make use of these symmetries to obtain alternative forms for the expression for the cost functional J given in (4.15). These alternative, but equivalent, forms may also be obtained using slight modifications of the evaluation technique.

The first modification relates to obtaining the required separation of the integrand into two parts. With $z \equiv e^{-sh}$, $A \equiv A(s)$, $\bar{A} \equiv A(-s)$, etc., the integrand of J may be written

$$E(s)E(-s) = \left(\frac{B + Dz}{A + Cz}\right)\left(\frac{\bar{B} + \bar{D}z^{-1}}{\bar{A} + \bar{C}z^{-1}}\right). \tag{4.20}$$

With the use of partial fractions, it is now possible to write

$$\left(\frac{B + Dz}{A + Cz}\right)\left(\frac{\bar{B} + \bar{D}z^{-1}}{\bar{A} + \bar{C}z^{-1}}\right) \equiv \frac{\alpha + \beta z}{A + Cz} + \frac{\gamma + \delta z^{-1}}{\bar{A} + \bar{C}z^{-1}} \tag{4.21}$$

in which α, β, γ and δ are independent of z. Cross-multiplying and equating powers of z then yields the system of equations

$$\begin{aligned} \alpha\bar{A} + \beta\bar{C} + \gamma A + \delta C &= B\bar{B} + D\bar{D} \\ \beta\bar{A} + \gamma C &= \bar{B}D \\ \alpha\bar{C} + \delta A &= B\bar{D}. \end{aligned} \tag{4.22}$$

Clearly the solution of this system is not unique since there are four unknowns and only three equations. Consequently, the particular separation used in (4.13) is not unique, although the value of the cost functional J is, of course, unique. The one other solution to be considered here is that given by

$$\left.\begin{aligned} \alpha &= \frac{A(B\bar{B} + D\bar{D}) - 2BC\bar{D}}{2(A\bar{A} - C\bar{C})}, \quad \gamma = \bar{\alpha} \\ \beta &= \frac{2AD\bar{B} - C(B\bar{B} + D\bar{D})}{2(A\bar{A} - C\bar{C})}, \quad \delta = \bar{\beta}. \end{aligned}\right\} \tag{4.23}$$

This particular solution has the advantage that the symmetry is retained and, in this sense, is analogous to the separation used in the delay-free case.

The second modification relates to the deformation of the contour from $-i\infty$ to $i\infty$. Clearly, it is only necessary to deform the contour so as to avoid passing through any new poles that might arise on the imaginary axis, as was the case in the example considered. Any further deformation is purely for the sake of convenience, as in the general case above where the contour was deformed so that all the new poles would

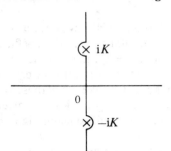

Fig. 4.2. The symmetrically deformed contour.

be on one side of the contour. If the separation given by (4.21) with (4.23) is used, it is more sensible to deform the contour so as to retain the symmetry. Consequently, in this case the contour is deformed to the left of any solution of (4.14) lying on the positive imaginary axis and to the right of any solution on the negative imaginary axis. This is shown in Fig. 4.2 for the case considered in the example. Then, the contour for the first part of the integral is closed on the right and for the second part on the left. The resulting expression for the cost functional J is

$$J = -\sum_{\text{r.h.p}} \text{res} \left\{ \frac{[A(B\bar{B} + D\bar{D}) - 2BC\bar{D}] + [2AD\bar{B} - C(B\bar{B} + D\bar{D})]e^{-sh}}{2(A\bar{A} - C\bar{C})(A + C e^{-sh})} \right\}$$

$$+ \sum_{\text{l.h.p}} \text{res} \left\{ \frac{[\bar{A}(B\bar{B} + D\bar{D}) - 2\overline{BCD}] + [2\bar{A}DB - \bar{C}(B\bar{B} + D\bar{D})]e^{sh}}{2(A\bar{A} - C\bar{C})(\bar{A} + \bar{C} e^{sh})} \right\} \quad (4.24)$$

in which the first summation is taken over the solution of (4.14) lying in the right half-plane or on the positive imaginary axis and the second summation over those in the left half plane or on the negative imaginary axis. However, by symmetry the two sets of residues are equal and opposite and so J may be written

$$J = -\sum_{\text{r.h.p}} \text{res} \left\{ \frac{[A(B\bar{B} + D\bar{D}) - 2BC\bar{D}] + [2AD\bar{B} - C(B\bar{B} + D\bar{D})]e^{-sh}}{(A\bar{A} - C\bar{C})(A + C e^{-sh})} \right\}. \quad (4.25)$$

One advantage of this form is that it does not require modification in the case when $A(s)$ and $C(s)$ possess a common root in the left half-plane, as (4.15) did. However, if both $A(s)$ and $C(s)$ possess a common factor s, the contour must also be deformed to the left of the origin and it may be shown that the contribution from the residue there is again that given by (4.19).

In many cases it is preferable to have an expression in terms of hyperbolic rather than exponential functions, since if the particular solution of (4.14) is pure imaginary, as was the case in the above example, such a form would immediately give J in terms of real quantities and, in particular, trigonometrical functions. For such purposes, it is convenient to introduce the even and odd parts of the polynomials $A(s)$, $B(s)$, $C(s)$ and $D(s)$, which will be denoted by

$$A_E = \tfrac{1}{2}(A + \bar{A}), \quad A_O = \tfrac{1}{2}(A - \bar{A}), \text{ etc.} \tag{4.26}$$

With the use of the identity

$$\frac{2A}{A + C e^{-sh}} = \frac{A + \bar{C} e^{sh}}{A_E + C_E \cosh sh - C_O \sinh sh}, \tag{4.27}$$

which is valid when $A\bar{A} = C\bar{C}$, it is then possible to show that

$$J = \sum_{\text{r.h.p}} \text{res} \frac{1}{(A\bar{A} - C\bar{C})(A_E + C_E \cosh sh - C_O \sinh sh)}$$
$$\times \{2(B_E D_E - B_O D_O)[C_O - A_O \cosh sh + A_E \sinh sh]$$
$$+ 2(B_O D_E - B_E D_O)[C_E + A_E \cosh sh - A_O \sinh sh]$$
$$+ (B_O^2 + D_O^2 - B_E^2 - D_E^2)[A_O - C_O \cosh sh + C_E \sinh sh]\}. \tag{4.28}$$

4.4 EXTENSION TO TWO COMMENSURATE DELAYS, h and $2h$

The method of evaluation given above is not restricted to systems involving a single time delay and may be extended to certain systems involving several delays. In this section we demonstrate this for systems involving two time delays, where the value of one is twice that of the other, i.e. they are commensurate. As before, a simple example will be considered first, and this will be that for which $E(s)$, the Laplace transform of the error $e(t)$, is given by

$$E(s) = \frac{1}{s + (1 - \alpha)e^{-sh} + \alpha e^{-2sh}} \tag{4.29}$$

in which α is a constant. In particular, if $\alpha = 0$ this reduces to (4.3), the single-delay example, and if $\alpha = 1$ this again reduces to (4.3) but with h replaced by $2h$. The cost functional J is then given by

$$J = \frac{1}{2\pi i} \int_{-i\infty}^{i\infty} E(s)E(-s)\,ds$$
$$= \frac{1}{2\pi i} \int_{-i\infty}^{i\infty} \frac{ds}{(s + (1 - \alpha)e^{-sh} + \alpha e^{-2sh})(-s + (1 - \alpha)e^{sh} + \alpha e^{2sh})}. \tag{4.30}$$

As with the single-delay case, the system is assumed to be stable and consequently all the poles of $E(s)$, or equivalently all the zeros of $(s + (1 - \alpha)e^{-sh} + \alpha e^{-2sh})$, lie in the left half of the complex s-plane.

The approach to be used is identical to that used previously in that the integrand is to be separated into two parts, then one part closed on the right and the other on the left. Consequently, in analogy with (4.21), with $z = e^{-sh}$, it is possible to write

$$\frac{1}{(s + (1 - \alpha)z + \alpha z^2)(-s + (1 - \alpha)z^{-1} + \alpha z^{-2})}$$

$$= \frac{L(s) + M(s)z + N(s)z^2}{s + (1-\alpha)z + \alpha z^2} + \frac{L(-s) + M(-s)z^{-1} + \alpha z^{-2}}{-s + (1-\alpha)z^{-1} + \alpha z^{-2}} \quad (4.31)$$

in which $L(s)$, $M(s)$, $N(s)$ are functions of s which are independent of h, or equivalently z, and use has been made of symmetry. Cross-multiplying and equating powers of z then yields the system of equations

$$-sL + (1-\alpha)M + \alpha N + s\bar{L} + (1-\alpha)\bar{M} + \alpha\bar{N} = 1$$
$$-sM + (1-\alpha)N + (1-\alpha)\bar{L} + \alpha\bar{M} = s\bar{M} + (1-\alpha)\bar{N} + (1-\alpha L + \alpha M = 0 \quad (4.32)$$
$$-sN + \alpha\bar{L} = s\bar{N} + \alpha L = 0$$

in which the notation $\bar{L} = L(-s)$ etc. has again been used. As before, the solution of this system is not unique, and consequently, in addition, it may be demanded that

$$N = \bar{N}. \quad (4.33)$$

The solution of the system is then found to be

$$L(s) = \frac{-s(s^2 + \alpha^2)}{\Delta(s)}, \quad M(s) = \frac{(1-\alpha)(s^2 + 2\alpha s - \alpha^2)}{2\Delta(s)}, \quad N(s) = \frac{\alpha(s^2 + \alpha^2)}{2\Delta(s)}$$

$$L(-s) = \frac{s(s^2 + \alpha^2)}{2\Delta(s)}, \quad M(-s) = \frac{(1-\alpha)(s^2 - 2\alpha s - \alpha^2)}{2\Delta(s)}, \quad N(-s) = N(s) \quad (4.34)$$

$$\Delta(s) = s^4 + [2\alpha^2 + (1-\alpha)^2]s^2 + (2\alpha - 1)\alpha^2.$$

Thus the cost functional J may now be written in the form

$$J = \frac{1}{2\pi i} \int_{-i\infty}^{i\infty} \left\{ \frac{-s(s^2 + \alpha^2) + (1-\alpha)(s^2 + 2\alpha s - \alpha^2)e^{-sh} + \alpha(s^2 + \alpha^2)e^{-2sh}}{s + (1-\alpha)e^{-sh} + \alpha e^{-2sh}} \right.$$
$$\left. + \frac{s(s^2 + \alpha^2) + (1-\alpha)(s^2 - 2\alpha s - \alpha^2)e^{sh} + \alpha(s^2 + \alpha^2)e^{2sh}}{-s + (1-\alpha)e^{sh} + \alpha e^{2sh}} \right\} \frac{ds}{2\Delta(s)}. \quad (4.35)$$

Following the procedure of indenting the contour from $-i\infty$ to $i\infty$ to the left of any zero of $\Delta(s)$ lying on the positive imaginary axis and to the right of any one the negative imaginary axis (as shown in Fig. 4.2), the contour may then be closed on the right for the first part of the integrand and on the left for the second. The contribution from these two parts will be identical by symmetry and consequently

$$J = -\sum_{\text{r.h.p}} \text{res} \left\{ \frac{-s(s^2 + \alpha^2) + (1-\alpha)(s^2 + 2\alpha s - \alpha^2)e^{-sh} + \alpha(s^2 + \alpha^2)e^{-2sh}}{\Delta(s)(s + (1-\alpha)e^{-sh} + \alpha e^{-2sh})} \right\} \quad (4.36)$$

in which the sum is taken over only those zeros of $\Delta(s)$ either in the right half of the complex s-plane or on the positive imaginary axis. It is also worth noting that the contributions from the semi-circles at infinity are indeed zero.

For this particular example the equation $\Delta(s) = 0$ is a quadratic in s^2 and the solutions may be written

Sec. 4.4] Extension to two commensurate delays, h and $2h$

$$2s^2 = -2\alpha^2 - (1-\alpha)^2 \pm |1-\alpha|(8\alpha^2 + (1-\alpha)^2)^{1/2} \tag{4.37}$$

from which it may be shown that the solutions for s^2 are real and negative for $\alpha > \frac{1}{2}$ and for $0 \neq \alpha < \frac{1}{2}$ are real with one positive and one negative. Thus the four roots for s are either pure imaginary or two are pure imaginary and two are real. For the former, the summation in (4.36) is therefore taken over the two of these lying on the positive imaginary axis and for the latter it is over the root on the positive imaginary axis and the root on the positive real axis. Moreover it turns out that the form of J given in (4.26) may be considerably simplified. This is because, at a zero of $\Delta(s)$, both the numerator and the deonominator of (4.26) factorize. To be more precise, at a zero of $\Delta(s)$,

$$s + (1-\alpha)e^{-sh} + \alpha e^{-2sh}$$

$$= \frac{\{(s^2+\alpha^2) + (1-\alpha)(s+\alpha)e^{-sh}\}\{s(s^2+\alpha^2) - \alpha(1-\alpha)(s-\alpha)e^{-sh}\}}{(s^2+\alpha^2)^2}$$

$$-s(s^2+\alpha^2) + (1-\alpha)(s^2+2\alpha s - \alpha^2)e^{-sh} + \alpha(s^2+\alpha^2)e^{-2sh} \tag{4.38}$$

$$= \frac{-\{(s^2+\alpha^2) - (1-\alpha)(s+\alpha)e^{-sh}\}\{s(s^2+\alpha^2) - \alpha(1-\alpha)(s-\alpha)e^{-sh}\}}{(s^2+\alpha^2)}$$

as may be verified by multiplying out and using the fact that $\Delta(s) = 0$. Since these contain a common factor, which will cancel, (4.36) reduces to

$$J = \sum_{\text{r.h.p}} \text{res} \frac{(s^2+\alpha^2)}{\Delta(s)} \left\{ \frac{(s^2+\alpha^2) - (1-\alpha)(s+\alpha)e^{-sh}}{(s^2+\alpha^2) + (1-\alpha)(s+\alpha)e^{-sh}} \right\}. \tag{4.39}$$

At first it might be thought that this factorization and consequent simplification is a result of the particular example chosen. On the contrary it will be shown that such simplifications always occur and a modification of the evaluation procedure, suitable for several commensurate delays, will be given which enables this form of the cost functional J to be obtained more directly. However, before leaving the above example and considering the general case of two commensurate delays, and systems containing more than two commensurate delays, it is worth confirming that (4.39) is consistent with the result for the single delay. The first check is obtained by setting $\alpha = 0$, when

$$J = \sum_{\text{r.h.p}} \text{res} \frac{s^2}{(s^4+s^2)} \left\{ \frac{s^2 - s e^{-sh}}{s^2 + s e^{-sh}} \right\}$$

$$= \text{res}_{s=i} \frac{(s-e^{-sh})}{(s^2+1)(s+e^{-sh})} = \frac{i - e^{-ih}}{2i(i+e^{-ih})} = \frac{\cos h}{2(1-\sin h)}, \tag{4.40}$$

which is in agreement with (4.8). To confirm the result for $\alpha = 1$, it is more convenient to use (4.36), which reduces to

$$J = -\sum_{\text{r.h.p}} \text{res}\left\{\frac{-s(s^2+1)+(s^2+1)e^{-2sh}}{(s^2+1)^2[s+e^{-2sh}]}\right\}$$

$$= -\operatorname*{res}_{s=i} \frac{(s-e^{-2sh})}{(s^2+1)(s+e^{-2sh})} = \frac{\cos 2h}{2(1-\sin 2h)}, \tag{4.41}$$

which agrees with (4.40) with h replaced by $2h$.

In general, for systems containing two commensurate delays h and $2h$, $E(s)$ will be of the form

$$E(s) = \frac{B(s) + D(s)e^{-sh} + G(s)e^{-2sh}}{A(s) + C(s)e^{-sh} + F(s)e^{-2sh}} \tag{4.42}$$

in which $A(s)$ to $G(s)$ are polynomials in s of finite degree and with real coefficients. As was seen when considering the single-delay case, minor modifications may be required if the polynomials possess a common factor, in particular $s = 0$. Consequently, in what follows it will be assumed that such common factors are not present. However, it should be emphasized that this is purely for convenience. The methods apply equally well in such cases but the answers may require modification.

Following the above procedure, the aim is to find functions $L(s)$, $M(s)$ and $N(s)$ such that

$$E(s)E(-s) = \frac{L(s) + M(s)z + N(s)z^2}{A(s) + C(s)z + F(s)z^2} + \frac{L(-s) + M(-s)z^{-1} + N(-s)z^{-2}}{A(-s) + C(-s)z^{-1} + F(-s)z^{-2}} \tag{4.43}$$

in which $z = e^{-sh}$ as before. Cross-multiplying and equating powers of z then yields the system of equations

$$L\bar{A} + \bar{L}A + M\bar{C} + \bar{M}C + N\bar{F} + \bar{N}F = B\bar{B} + D\bar{D} + G\bar{G}$$

$$M\bar{A} + N\bar{C} + \bar{L}C + \bar{M}F = \bar{B}D + \bar{D}G$$

$$\bar{M}A + \bar{N}C + L\bar{C} + M\bar{F} = B\bar{D} + D\bar{G} \tag{4.44}$$

$$N\bar{A} + \bar{L}F = \bar{B}G$$

$$\bar{N}A + L\bar{F} = B\bar{G}$$

using the previous notation. Again the solution of this sytem is not unique; however, in analogy with (4.34), one solution is

$$2\Delta L = 2\{(\bar{A}C - \bar{C}F)C - (A\bar{A} - F\bar{F})\}F)B\bar{G} - 2(\bar{A}C - \bar{C}F)A(B\bar{D} + D\bar{G})$$
$$+ (A\bar{A} - F\bar{F})A(B\bar{B} + D\bar{D} + G\bar{G})$$

$$2\Delta M = 2(A\bar{A} - F\bar{F})A(\bar{B}D + \bar{D}G) - 2(A\bar{A} - F\bar{F})F(B\bar{D} + D\bar{G}) - 2(A\bar{C} - C\bar{F})A\bar{B}G$$
$$+ 2(\bar{A}C - \bar{C}F)F B\bar{G} - [(\bar{A}C - \bar{C}F)A - (A\bar{C} - C\bar{F})F](B\bar{B} + D\bar{D} + G\bar{G})$$

$$2\Delta N = 2\{(A\bar{A} - F\bar{F})A - (A\bar{C} - C\bar{F})C\}\bar{B}G + 2(A\bar{C} - C\bar{F})F(\bar{B}D + \bar{D}G)$$
$$- (A\bar{A} - F\bar{F})F(B\bar{B} + D\bar{D} + G\bar{G})$$

$$\tag{4.45}$$

where
$$\Delta = (A\bar{A} - F\bar{F})^2 - (A\bar{C} - C\bar{F})(\bar{A}C - \bar{C}F). \tag{4.46}$$

The cost functional J is then given by

$$J = -2 \sum_{\text{r.h.p}} \text{res} \left\{ \frac{L(s) + M(s)z + N(s)z^2}{A(s) + C(s)z + F(s)z^2} \right\} \tag{4.47}$$

where the sum is taken over those zeros of $\Delta(s)$ either in the right half of the complex s-plane or on the positive imaginary axis. As before it is possible to show that both the numerator and the denominator in (4.47) possess a common factor and it may be shown that (4.47) reduces to

$$J = -\sum_{\text{r.h.p}} \text{res} \frac{1}{\Delta\{(A\bar{A} - F\bar{F}) + (\bar{A}C - \bar{C}F)z\}} \left\{ -2(\bar{A}C - \bar{C}F)(A\bar{A} - F\bar{F})(B\bar{D} + D\bar{G}) \right.$$
$$+ 2(\bar{A}C - \bar{C}F)^2 B\bar{G} + 2(A\bar{A} - F\bar{F})^2(\bar{B}D + \bar{D}G)z - 2(A\bar{A} - F\bar{F})(A\bar{C} - C\bar{F})\bar{B}Gz$$
$$\left. + (A\bar{A} - F\bar{F})[(A\bar{A} - \bar{F}F) - (\bar{A}C - \bar{C}F)z](B\bar{B} + D\bar{D} + G\bar{G}) \right\} \tag{4.48}$$

where the summation is taken over those zeros of $\Delta(s)$ either in the right half of the complex s-plane or on the positive imaginary axis.

This factorization may be obtained more directly using an extension of the original method. To demonstrate this, the example in which

$$J = \frac{1}{2\pi i} \int_{-i\infty}^{i\infty} \frac{ds}{(s + (1-\alpha)e^{-sh} + \alpha e^{-2sh})(-s + (1-\alpha)e^{sh} + \alpha e^{2sh})} \tag{4.49}$$

is again considered. Closing the contour on the right then gives

$$J = -\sum_{(1)} \text{res} \left\{ \frac{1}{(s + (1-\alpha)e^{-sh} + \alpha e^{-2sh})(-s + (1-\alpha)e^{sh} + \alpha e^{2sh})} \right\} \tag{4.50}$$

where the summation is taken over the solutions of

$$(1): -s + (1-\alpha)e^{sh} + \alpha e^{2sh} = 0 \tag{4.51}$$

all of which are assumed to lie in the right half of the complex s-plane. However, by rearranging (4.51), it follows that at such a point

$$e^{-2sh} = \frac{\alpha + (1-\alpha)e^{-sh}}{s} \tag{4.52}$$

and hence (4.50) may be replaced by

$$J = -\sum_{(1)} \text{res} \left\{ \frac{s}{((s^2 + \alpha^2) + (1-\alpha)(s+\alpha)e^{-sh})(-s + (1-\alpha)e^{sh} + \alpha e^{2sh})} \right\}. \tag{4.53}$$

Since the integral round the contour at infinity for an integrand such as that shown in (4.53) is zero, the summation over *all* the residues must be zero and consequently

$$J = \sum_{(2)} \text{res} \left\{ \frac{s}{((s^2 + \alpha^2) + (1 - \alpha)(s + \alpha)e^{-sh})(-s + (1 - \alpha)e^{sh} + \alpha e^{2sh})} \right\} \quad (4.54)$$

where the summation is now taken over the solutions of

$$(2): (s^2 + \alpha^2) + (1 - \alpha)(s + \alpha)e^{-sh} = 0. \quad (4.55)$$

However, by rearranging (4.55), it follows that at such a point

$$e^{sh} = -\frac{(1 - \alpha)(s + \alpha)}{(s^2 + \alpha^2)} \quad (4.56)$$

and hence (4.54) may be replaced by

$$J = \sum_{(2)} \text{res} \left\{ -\frac{(s^2 + \alpha^2)^2}{((s^2 + \alpha^2) + (1 - \alpha)(s + \alpha)e^{-sh})\Delta(s)} \right\} \quad (4.57)$$

where

$$\Delta(s) = s^4 + [2\alpha^2 + (1 - \alpha)^2]s^2 + (2\alpha - 1)\alpha^2 \quad (4.58)$$

as before. Again the summation over all the residues must be zero and consequently

$$J = \sum_{(3)} \text{res} \frac{1}{\Delta(s)} \left\{ \frac{(s^2 + \alpha^2)^2}{(s^2 + \alpha^2) + (1 - \alpha)(s + \alpha)e^{-sh}} \right\} \quad (4.59)$$

where the summation is taken over the zeros of $\Delta(s)$.

To show that this is equivalent to (4.39), use is made of the fact that the zeros of $\Delta(s)$ occur in equal and opposite pairs. This enables (4.50) to be written

$$J = \sum_{\text{r.h.p}} \text{res} \frac{1}{\Delta(s)}$$

$$\left\{ \frac{(s^2 + \alpha^2)^2}{(s^2 + \alpha^2) + (1 - \alpha)(s + \alpha)e^{-sh}} - \frac{(s^2 + \alpha^2)^2}{(s^2 + \alpha^2) + (1 - \alpha)(-s + \alpha)e^{sh}} \right\} \quad (4.60)$$

where the summation is now taken over only those zeros of $\Delta(s)$ lying in the right half-plane or on the positive imaginary axis, and the second term in (4.60) was obtained by replacing s by $-s$ in (4.59). Rearrangement of (4.60) and use of the identity $\Delta(s) = 0$ or equivalently

$$(s^2 + \alpha^2)^2 = (1 - \alpha)^2(-s + \alpha)(s + \alpha) \quad (4.61)$$

then yields

$$J = \sum_{\text{r.h.p}} \text{res} \frac{(s^2 + \alpha^2)}{\Delta(s)} \left\{ \frac{(s^2 + \alpha^2) - (1 - \alpha)(s + \alpha)e^{-sh}}{(s^2 + \alpha^2) + (1 - \alpha)(s + \alpha)e^{-sh}} \right\}, \quad (4.62)$$

which is precisely (4.39).

4.5 SEVERAL COMMENSURATE DELAYS

The above approach of reducing the number of exponential terms within a given expression and then changing to calculating residues over the reduced expression may be used equally well for systems involving multiple commensurate delays. It will now be shown that it is in fact possible to obtain a recursive scheme, analogous to that given by Åström (1970) for the delay-free case, for such systems. For a general system involving n commensurate delays, $E(s)$, the Laplace transform of $e(t)$, will be of the form

$$E(s) = \frac{B_0(s) + B_1(s)z + \ldots + B_n(s)z^n}{A_0(s) + A_1(s)z + \ldots + A_n(s)z^n} \quad (4.63)$$

in which $z = e^{-sh}$ and $A_0(s), A_1(s), \ldots, A_n(s)$ and $B_0(s), B_1(s), \ldots, B_n(s)$ are polynomials in s with real coefficients and in analogy with our earlier assumption $A_0(s)$ is taken to dominate. The cost functional J is then given by

$$J = \frac{1}{2\pi i} \int_{-i\infty}^{i\infty} E(s)E(-s)\,ds$$

$$= \frac{1}{2\pi i} \int_{-i\infty}^{i\infty} \frac{\left(\sum_0^n B_r z^r\right)\left(\sum_0^n \bar{B}_r z^{-r}\right)}{\left(\sum_0^n A_r z^r\right)\left(\sum_0^n \bar{A}_r z^{-r}\right)}\,ds, \quad (4.64)$$

again making use of the shorthand notation $A_r \equiv A_r(s)$, $\bar{A}_r \equiv A_r(-s)$, etc.

The system is again assumed stable so that all the poles of $E(s)$ lie in the left half of the complex s-plane or equivalently all the zeros of

$$\left(\sum_0^n A_r z^r\right).$$

The first step is to close the contour in the left half-plane and here, as throughout, it will be assumed that the contribution from the contour at infinity is zero. (4.64) then yields

$$J = \sum_A \mathrm{res}\, \frac{\left(\sum_0^n B_r z^r\right)\left(\sum_0^n \bar{B}_r z^{-r}\right)}{\left(\sum_0^n A_r z^r\right)\left(\sum_0^n \bar{A}_r z^{-r}\right)} \quad (4.65)$$

where the summation is taken over the solutions of

$$A: \sum_0^n A_r z^r \equiv A_0 + A_1 z + \ldots + A_n z^n = 0. \quad (4.66)$$

Now at such a solution, it is possible to solve (4.66) for z^{-n} and then replace the terms in z^{-n} in (4.65). To be more precise, at such a solution,

$$A_0 z^{-n} = -(A_n + A_{n-1} z^{-1} + \ldots + A_1 z^{-(n-1)}) \tag{4.67}$$

and so

$$\frac{\sum_0^n \bar{B}_r z^{-r}}{\sum_0^n \bar{A}_r z^{-r}}$$

$$= \frac{A_0(\bar{B}_0 + \bar{B}_1 z^{-1} + \ldots + \bar{B}_{n-1} z^{-(n-1)}) - \bar{B}_n(A_n + A_{n-1} z^{-1} + \ldots + A_1 z^{-(n-1)})}{A_0(\bar{A}_0 + \bar{A}_1 z^{-1} + \ldots + \bar{A}_{n-1} z^{-(n-1)}) - \bar{A}_n(A_n + A_{n-1} z^{-1} + \ldots + A_1 z^{-(n-1)})}$$

$$= \frac{\sum_0^{n-1} (A_0 \bar{B}_r - \bar{B}_n A_{n-r}) z^{-r}}{\sum_0^{n-1} \bar{C}_r z^{-r}} \tag{4.68}$$

where C_r is defined by

$$C_r = \bar{A}_0 A_r - A_n \bar{A}_{n-r}. \tag{4.69}$$

Substituting the left-hand side of (4.68) into (4.65) then gives

$$J = \sum_A \operatorname{res} \frac{\left(\sum_0^n B_r z^r\right)}{\left(\sum_0^n A_r z^r\right)} \left\{ \frac{\sum_0^{n-1} (A_0 \bar{B}_r - \bar{B}_n A_{n-r}) z^{-r}}{\sum_0^{n-1} \bar{C}_r z^{-r}} \right\}. \tag{4.70}$$

Assuming that the contribution from the circle at infinity is, as always, zero, the summation of *all* the residues is zero and hence

$$J = -\sum_{\bar{C}} \operatorname{res} \frac{\left(\sum_0^n B_r z^r\right)}{\left(\sum_0^n A_r z^r\right)} \left\{ \frac{\sum_0^{n-1} (A_0 \bar{B}_r - \bar{B}_n A_{n-r}) z^{-r}}{\sum_0^{n-1} \bar{C}_r z^{-r}} \right\} \tag{4.71}$$

where the summation is now taken over the solutions of

$$\bar{C} : \sum_0^{n-1} \bar{C}_r z^{-r} \equiv \bar{C}_0 + \bar{C}_1 z^{-1} + \ldots + \bar{C}_{n-1} z^{-(n-1)} = 0. \tag{4.72}$$

Moreover, at such a solution

$$\bar{C}_0 z^{n-1} = -(\bar{C}_{n-1} + \bar{C}_{n-2} z + \ldots + \bar{C}_1 z^{n-2}) \tag{4.73}$$

and so

$$\frac{\sum_0^n B_r z^r}{\sum_0^n A_r z^r}$$

$$= \frac{\bar{C}_0(B_0 + B_1 z + \ldots + B_{n-1}z^{(n-1)}) - B_n z(\bar{C}_{n-1} + \bar{C}_{n-2}z + \ldots + \bar{C}_1 z^{(n-2)})}{\bar{C}_0(A_0 + A_1 z + \ldots + A_{n-1}z^{(n-1)}) - A_n z(\bar{C}_{n-1} + \bar{C}_{n-2}z + \ldots + \bar{C}_1 z^{(n-2)})}$$

$$= \frac{\sum_0^{n-1} (\bar{C}_0 B_r - B_n \bar{C}_{n-r}) z^r}{A_0 \sum_0^{n-1} C_r z^r} \tag{4.74}$$

since $\bar{C}_n = 0$ and

$$\bar{C}_0 A_r - A_n \bar{C}_{n-r} = (A_0 \bar{A}_0 - \bar{A}_n a_n) A_r - A_n(A_0 \bar{A}_{n-r} - \bar{A}_n A_r)$$
$$= A_0(\bar{A}_0 A_r - A_n \bar{A}_{n-r}) = A_0 C_r. \tag{4.75}$$

Substitution of (4.74) into (4.71) then gives

$$J = -\sum_{\bar{C}} \mathrm{res}\, \frac{\left\{\sum_0^{n-1} (\bar{C}_0 B_r - B_n \bar{C}_{n-r}) z^r\right\} \left\{\sum_0^{n-1} (A_0 \bar{B}_r - \bar{B}_n A_{n-r}) z^{-r}\right\}}{A_0 \left(\sum_0^{n-1} C_r z^r\right)\left(\sum_0^{n-1} \bar{C}_r z^{-r}\right)}. \tag{4.76}$$

However, from (4.75) with r replaced by $(n - r)$

$$\bar{C}_0 A_{n-r} - A_n \bar{C}_r = A_0 C_{n-r} \tag{4.77}$$

and so

$$\bar{C}_0 \sum_0^{n-1} A_{n-r} z^{-r} - A_n \sum_0^{n-1} \bar{C}_r z^{-r} = A_0 \sum_0^{n-1} C_{n-r} z^{-r}. \tag{4.78}$$

Thus, at a residue,

$$\bar{C}_0 \sum_0^{n-1} A_{n-r} z^{-r} = A_0 \sum_0^{n-1} C_{n-r} z^{-r} \tag{4.79}$$

and hence

$$A_0 \sum_0^{n-1} \bar{B}_r z^{-r} - \bar{B}_n \sum_0^{n-1} A_{n-r} z^{-r} = \frac{A_0}{\bar{C}_0}\left\{\bar{C}_0 \sum_0^{n-1} \bar{B}_r z^{-r} - \bar{B}_n \sum_0^{n-1} C_{n-r} z^{-r}\right\} \tag{4.80}$$

and (4.76) may be written

$$J = -\sum_{\bar{C}} \mathrm{res}\, \frac{1}{C_0} \frac{\left(\sum_0^{n-1} D_r z^r\right)\left(\sum_0^{n-1} \bar{D}_r z^{-r}\right)}{\left(\sum_0^{n-1} C_r z^r\right)\left(\sum_0^{n-1} \bar{C}_r z^{-r}\right)} \tag{4.81}$$

in which D_r is defined by

$$D_r = C_0 B_r - B_n \bar{C}_{n-r} \tag{4.82}$$

and use has been made of the fact that $\bar{C}_0 = C_0$ and is hence symmetrical in s. Similarly the rest of the expression in (4.81) is also symmetrical and consequently the residues of C_0 will give a zero contribution. Thus since the sum of *all* the residues is zero it follows that

$$J = -\sum_C \mathrm{res}\, \frac{1}{C_0} \frac{\left(\sum_0^{n-1} D_r z^r\right)\left(\sum_0^{n-1} \bar{D}_r z^{-r}\right)}{\left(\sum_0^{n-1} C_r z^r\right)\left(\sum_0^{n-1} \bar{C}_r z^{-r}\right)} \tag{4.83}$$

where the summation is now taken over the solutions of

$$C: \sum_0^{n-1} C_r z^r \equiv C_0 + C_1 z + \ldots + C_{n-1} z^{n-1} = 0. \tag{4.84}$$

Apart from the factor $(1/C_0)$, the expression in (4.83) is of the same form as that in (4.65) but involving one less power of z. With appropriate modifications to take account of this additional factor the above procedure may be repeated. The first stage leads to the analogue of (4.70), namely

$$J = \sum_C \mathrm{res}\, \frac{1}{C_0} \frac{\left(\sum_0^{n-1} D_r z^r\right)}{\left(\sum_0^{n-1} C_r z^r\right)} \left\{ \frac{\sum_0^{n-2} (C_0 \bar{D}_r - \bar{D}_{n-1} C_{n-1-r}) z^{-r}}{\sum_0^{n-2} \bar{E}_r z^{-r}} \right\} \tag{4.85}$$

in which E_r is defined by

$$E_r = \bar{C}_0 C_r - C_{n-1} \bar{C}_{n-1-r}. \tag{4.86}$$

The next step is to change to a summation over the solutions of

$$\bar{E}: \sum_0^{n-2} \bar{E}_r z^{-r} = \bar{E}_0 + \bar{E}_1 z^{-1} + \ldots + \bar{E}_{n-2} z^{-(n-2)} = 0. \tag{4.87}$$

However, consideration must also be given to any residues that arise from the zeros of C_0. At such a zero,

$$\frac{\left(\sum_0^{n-1} D_r z^r\right)}{\left(\sum_0^{n-1} C_r z^r\right)} \left\{ \frac{\sum_0^{n-2} (C_0 \bar{D}_r - \bar{D}_{n-1} C_{n-1-r}) z^{-r}}{\sum_0^{n-2} \bar{E}_r z^{-r}} \right\}$$

Sec. 4.5] Several commensurate delays

$$= \frac{\left(\sum_{0}^{n-1} D_r z^r\right)}{\left(\sum_{0}^{n-1} C_r z^r\right)} \frac{\bar{D}_{n-1}}{\bar{C}_{n-1}}$$

$$= \frac{B_n \left(\sum_{0}^{n-1} \bar{C}_{n-r} z^r\right)}{\left(\sum_{0}^{n-1} C_r z^r\right)} \frac{\bar{B}_n C_1}{\bar{C}_{n-1}}$$

$$= \frac{B_n A_0 \left(\sum_{0}^{n-1} C_r z^r\right)}{\left(\sum_{0}^{n-1} C_r z^r\right)} \frac{\bar{B}_n C_1}{A_0 C_1} = B_n \bar{B}_n \tag{4.88}$$

making use of (4.75) and (4.82) with $C_0 = \bar{C}_0 = 0$. Since this is symmetrical in s it follows that the total contribution from all the residues at the zeros of C_0 is zero and hence

$$J = -\sum_{\bar{E}} \text{res} \, \frac{1}{C_0} \frac{\left(\sum_{0}^{n-1} D_r z^r\right)}{\left(\sum_{0}^{n-1} C_r z^r\right)} \left\{ \frac{\sum_{0}^{n-2} (C_0 \bar{D}_r - \bar{D}_{n-1} C_{n-1-r}) z^{-r}}{\sum_{0}^{n-2} \bar{E}_r z^{-r}} \right\} \tag{4.89}$$

where the summation is taken over the solutions of $\bar{E}$, as defined in (4.87). Following the procedure from (4.71) to (4.81) then yields

$$J = -\sum_{\bar{E}} \text{res} \, \frac{1}{C_0 E_0}$$

$$\frac{\left\{\sum_{0}^{n-2} (E_0 D_r - D_{n-1} \bar{E}_{n-1-r}) z^r\right\} \left\{\sum_{0}^{n-2} (E_0 \bar{D}_r - \bar{D}_{n-1} E_{n-1-r}) z^{-r}\right\}}{\left(\sum_{0}^{n-2} E_r z^r\right) \left(\sum_{0}^{n-2} \bar{E}_r z^{-r}\right)}. \tag{4.90}$$

Now

$$E_0 D_r - D_{n-1} \bar{E}_{n-1-r} = C_0(\bar{C}_0 D_r - \bar{C}_{n-1-r} D_{n-1}) - \bar{C}_{n-1}(C_{n-1} D_r - C_r D_{n-1}) \tag{4.91}$$

from (4.86) and

$$C_{n-1} D_r = C_0(C_{n-1} B_r - C_r B_{n-1}) - B_n(C_{n-1} \bar{C}_{n-r} - \bar{C}_1 C_r) \tag{4.92}$$

from (4.82) and

$$C_{n-1} \bar{C}_{n-r} - \bar{C}_1 C_r = C_0(A_{n-1} \bar{A}_{n-r} - \bar{A}_1 A_r) \tag{4.93}$$

from (4.69) and (4.75). Thus $(E_0 D_r - D_n \bar{E}_{n-1-r})$ has a factor C_0 and so with F_r defined by

$$F_r = \frac{E_0 D_r - D_{n-1} \bar{E}_{n-1-r}}{C_0} \tag{4.94}$$

(4.90) becomes

$$J = -\sum_E \operatorname{res} \frac{C_0}{E_0} \frac{\left(\sum_0^{n-2} F_r z^r\right)\left(\sum_0^{n-2} \bar{F}_r z^{-r}\right)}{\left(\sum_0^{n-2} E_r z^r\right)\left(\sum_0^{n-2} \bar{E}_r z^{-r}\right)} \tag{4.95}$$

$$= \sum_E \operatorname{res} \frac{C_0}{E_0} \frac{\left(\sum_0^{n-2} F_r z^r\right)\left(\sum_0^{n-2} \bar{F}_r z^{-r}\right)}{\left(\sum_0^{n-2} E_r z^r\right)\left(\sum_0^{n-2} \bar{E}_r z^{-r}\right)}$$

where the summation is taken over the solutions of

$$E: \sum_0^{n-2} E_r z^r \equiv E_0 + E_1 z + \ldots + E_{n-2} z^{n-2} \tag{4.96}$$

and again this is possible since there will be no contribution from the zeros of E_0 by symmetry.

This entire process may be repeated again to yield

$$J = \sum_G \operatorname{res} \frac{E_0}{G_0} \frac{\left(\sum_0^{n-3} H_r z^r\right)\left(\sum_0^{n-3} \bar{H}_r z^{-r}\right)}{\left(\sum_0^{n-3} G_r z^r\right)\left(\sum_0^{n-3} \bar{G}_r z^{-r}\right)} \tag{4.97}$$

where

$$G_r = \frac{E_0 E_r - E_{n-2} \bar{E}_{n-2-r}}{C_0}, \quad H_r = \frac{G_0 F_r - F_{n-2} \bar{G}_{n-2-r}}{E_0} \tag{4.98}$$

and again it may be shown that $(G_0 F_r - F_{n-2} \bar{G}_{n-2-r})$ has a factor E_0 and that $(E_0 E_r - E_{n-2} \bar{E}_{n-2-r})$ has a factor C_0. This is the recurrence formulation analogous to that given by Åström (1970) and may be repeated as many times as required. Further details are to be found in Walton and Marshall (1987b). Of course, at the final stage, the procedure is terminated at the half-way stage when one of the polynomials (in z) in the denominator becomes independent of z. For example, for the case of quadratics in z, $n = 2$ and (4.63) becomes

$$E(s) = \frac{B_0(s) + B_1(s)z + B_2(s)z^2}{A_0(s) + A_1(s)z + A_2(s)z^2} \tag{4.99}$$

then

$$C_0 = A_0\bar{A} - A_2\bar{A}_2, \quad C_1 = \bar{A}_0 A_1 - A_2\bar{A}_1$$
$$D_0 = C_0 B_0, \quad D_1 = C_0 B_1 - B_2 \bar{C}_1 \tag{4.100}$$
$$E_0 = C_0 \bar{C}_0 - C_1 \bar{C}_1$$

and the process is terminated at (4.89) which, in this case, becomes

$$J = -\sum_{E_0} \operatorname{res} \frac{(C_0 \bar{D}_0 - \bar{D}_1 C_1)}{C_0 E_0} \left\{ \frac{D_0 + D_1 z}{C_0 + C_1 z} \right\} \tag{4.101}$$

where the summation is taken over the zeros of E_0, which is a finite polynomial in s, or, alternatively, in terms of the original polynomials, at the solutions of

$$E_0 = C_0 \bar{C}_0 - C_1 \bar{C}_1 = (A_0 \bar{A}_0 - A_2 \bar{A}_2)^2 - (\bar{A}_0 A_1 - A_2 \bar{A}_1)(A_0 \bar{A}_1 - \bar{A}_2 A_1) = 0$$
$$\tag{4.102}$$

which is seen to be consistent with (4.46) with the appropriate change in notation. Moreover it may be shown that (4.101) is equivalent to (4.48), as is to be expected.

On several occasions in this section it has been assumed that 'the contribution from the contour at infinity is zero'. In practice it is sufficient that the polynomial $A_0(s)$ dominates all the other polynomials in $E(s)$ in the sense that it has higher degree. Although this does not ensure that the above 'contribution' is zero it does ensure that if it is not then it will be cancelled by the contribution from the next contour at infinity.

4.6 EXTENSION TO WEIGHTED QUADRATIC CRITERIA

In this chapter the only cost functional to have been considered so far is that of the integral of square error, namely

$$J = \int_0^\infty e^2(t)\,dt \tag{4.103}$$

for which, by Parseval's theorem, there is the alternative representation

$$J = \frac{1}{2\pi i} \int_{-i\infty}^{i\infty} E(s)E(-s)\,ds \tag{4.104}$$

where $E(s)$ denotes the Laplace transform of the error $e(t)$. Moreover it is this latter representation which forms the basis for the evaluation technique of the chapter. It will now be shown that, in principle, this technique is equally applicable to more general cost functions of the form

$$J_{qm} = \int_0^\infty t^q \left(\frac{d^m e}{dt^m}\right)^2 dt \qquad (4.105)$$

in which q and m are non-negative integers.

Firstly for any two functions $f(t)$ and $g(t)$, suitably well behaved, with Laplace transforms $F(s)$ and $G(s)$, the application of Parseval's theorem yields

$$\int_0^\infty f(t)g(t)\,dt = \frac{1}{2\pi i}\int_{-i\infty}^{i\infty} F(s)G(-s)\,ds = \frac{1}{2\pi i}\int_{-i\infty}^{i\infty} F(-s)G(s)\,ds. \qquad (4.106)$$

It is then possible, by suitable choices of $f(t)$ and $g(t)$, for example

$$f(t) = t^q \frac{d^m e}{dt^m}, \quad g(t) = \frac{d^m e}{dt^m} \qquad (4.107)$$

to obtain $F(s)$ and $G(-s)$. The resulting integral in s will then be of a similar form to those previously considered and the methods presented there may be extended to such cases. This will be demonstrated by considering three examples and in all cases, for simplicity, attention will be restricted to examples involving a single time delay.

The first example to be considered is that in which $q = 0$ and $m = 1$. Thus

$$J_{01} = \int_0^\infty \left(\frac{de}{dt}\right)^2 dt. \qquad (4.108)$$

Moreover since

$$\mathscr{L}\left(\frac{de}{dt}\right) = sE(s) - e(0) \qquad (4.109)$$

it follows that if, for the most general form of $E(s)$ involving a single time delay

$$E(s) = \frac{B(s) + D(s)e^{-sh}}{A(s) + C(s)e^{-sh}} \qquad (4.110)$$

as in (4.91), then

$$sE(s) - e(0) = \frac{\tilde{B}(s) + \tilde{D}(s)e^{-sh}}{A(s) + C(s)e^{-sh}} \qquad (4.111)$$

where

$$\tilde{B}(s) = sB(s) - e(0)A(s), \quad \tilde{D}(s) = sD(s) - e(0)C(s). \qquad (4.112)$$

Since this is of the general form already considered, the method of evaluation is exactly as before and the value of the cost functional is given by (4.24) with B and D replaced respectively by $\tilde{B}$ and $\tilde{D}$. It is also worth remarking that J_{0m} for arbitrary positive values of m may be evaluated in an identical manner since it will also give rise to an integral of this general form. With such examples a little caution is recommended since it is wise to check that the polynomial $A(s)$ does still dominate. The second example to be considered is that in which $q = 1$ and $m = 0$. Thus

$$J_{10} = \int_0^\infty t e^2(t)\, dt. \tag{4.113}$$

Since

$$\mathcal{L}(t\, e(t)) = -\frac{dE(s)}{ds} \tag{4.114}$$

it follows from (4.106) that

$$J_{10} = -\frac{1}{2\pi i} \int_{-i\infty}^{i\infty} \frac{dE(s)}{ds} E(-s)\, ds. \tag{4.115}$$

To demonstrate how this integral may be evaluated, the simple example

$$E(s) = \frac{1}{s + K e^{-sh}} \tag{4.116}$$

is again considered and hence in this case

$$J_{10} = \frac{1}{2\pi i} \int_{-i\infty}^{i\infty} \frac{(1 - Kh e^{-sh})}{(s + K e^{-sh})^2} \frac{ds}{(-s + K e^{sh})}. \tag{4.117}$$

As before, the aim is to separate the integrand into two parts. As has already been seen, there are different, but equivalent, ways of achieving this. One way is to say that at a zero of $(-s + K e^{sh})$, $e^{-sh} = K/s$ and so

$$J_{10} = \frac{1}{2\pi i} \oint_{-i\infty}^{i\infty} \frac{ds}{(-s + K e^{sh})} \left\{ \frac{1 - Kh e^{-sh}}{(s + K e^{-sh})^2} - \frac{s(s - hK^2)}{(s^2 + K^2)^2} \right\}$$

$$+ \frac{1}{2\pi i} \oint_{-i\infty}^{i\infty} \frac{s(s - hK^2)\, ds}{(-s + K e^{sh})(s^2 + K^2)^2} \tag{4.118}$$

where the contour is deformed to the left of $s = \pm iK$. Then, closing the contour on the right for the first integral and on the left for the second yields

$$J_{10} = -\sum_{s = \pm iK} \operatorname{res} \frac{s(s - hK^2)}{(-s + K e^{sh})(s^2 + K^2)^2}$$

$$= \frac{1 + K^2 h^2 - Kh \cos Kh}{4K^2(1 - \sin Kh)} \tag{4.119}$$

since there is zero contribution from the first integral since there are no enclosed poles. However, some care must be taken when evaluating such integrals since double poles are present. We note that (4.119) agrees with (3.68).

The third example is that in which $q = 2$ and $m = 0$. Thus

$$J_{20} = \int_0^\infty t^2 e^2(t)\,dt \tag{4.120}$$

and retaining the symmetry and taking $f(t) = g(t) = te(t)$ in (4.106) it follows that

$$J_{20} = \frac{1}{2\pi i} \int_{-i\infty}^{i\infty} \frac{dE(s)}{ds} \frac{dE(-s)}{d(-s)}\,ds. \tag{4.121}$$

Again taking $E(s)$ to be of the form in (4.116),

$$J_{20} = \frac{1}{2\pi i} \int_{-i\infty}^{i\infty} \frac{1 - Khe^{-sh}}{(s + Ke^{-sh})^2} \frac{1 - Khe^{sh}}{(-s + Ke^{sh})^2}\,ds. \tag{4.122}$$

In this case it is more convenient to seek a symmetric separation, which is found to be

$$\frac{1 - Khe^{-sh}}{(s + Ke^{-sh})^2} \frac{1 - Khe^{sh}}{(-s + Ke^{sh})^2} = \frac{1}{2(s^2 + K^2)^3}$$

$$\left\{ \frac{s^2(s^2 - K^2)(1 + K^2 h^2) - 2(s^2 + K^2)[s(1 + h^2 K^2) + Kh(K^2 + s^2)]e^{-sh} - (s^2 - K^2)(1 + K^2 h^2)e^{-2sh}}{(s + Ke^{-sh})^2} \right.$$

$$\left. + \frac{s^2(s^2 - K^2)(1 + K^2 h^2) - 2(s^2 + K^2)[-s(1 + h^2 K^2) + Kh(K^2 + s^2)]e^{sh} - (s^2 - K^2)(1 + K^2 h^2)e^{2sh}}{(-s + Ke^{sh})^2} \right\}. \tag{4.123}$$

Deforming the contour as shown in Fig. 4.2 and closing the contour as before leads to

$$J = -\operatorname*{res}_{s = iK} \frac{1}{(s^2 + K^2)^3}$$

$$\left\{ \frac{s^2(s^2 - K^2)(1 + K^2 h^2) - 2(s^2 + K^2)[s(1 + h^2 K^2) + Kh(K^2 + s^2)]e^{-sh} - (s^2 - K^2)(1 + K^2 h^2)e^{-2sh}}{(s + Ke^{-sh})^2} \right\}$$

$$= \frac{(1 + K^2 h^2)}{8K^3(1 - \sin Kh)} \left\{ \frac{(2 - \sin Kh)\cos Kh}{1 - \sin Kh} - Kh + \frac{K^2 h^2 \cos Kh}{1 - \sin Kh} \right\}$$

$$-\frac{h}{2K^2(1 - \sin Kh)} \tag{4.124}$$

in agreement with (3.69).

The above examples demonstrate how the techniques presented in this chapter may be used to evaluate integrals of the type shown in (4.105). The additional complications that arise are that double poles may be present and that case must be taken when deriving the appropriate form of separation to ensure that the contributions at infinity will be zero. Also, as is always the case, any common factor in the denominator will require special attention. Further details on the above extensions are to be found in Walton, Ireland and Marshall (1986) in which the case of exponentially weighted criteria is also considered.

4.7 STORED FUNCTIONS, ARBITRARY INPUTS AND INITIAL CONDITIONS

In the context of systems with one delay, the present chapter has been restricted to those for which $E(s)$, the transform of error, is of the form given in (4.9), some justification for which is given in Chapter 2. In this section we look more closely at this but with particular reference to any function that may be stored in the delay, to an arbitrary input to the system or to arbitrary initial conditions. In such cases, does (4.9) still apply and, if not, is it possible to extend the method given earlier?

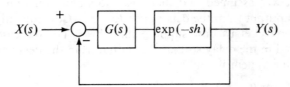

Fig. 4.3. Time-delay system with sub-plant $G(s)$.

To attempt to answer these questions, attention is restricted to the system depicted in Fig. 4.3, where $X(s)$ and $Y(s)$ denote respectively the Laplace transforms of the input $x(t)$ and the output $y(t)$. They are related by

$$Y = (X - Y)G e^{-sh} \tag{4.125}$$

and consequently $E(s)$, the Laplace transform of the error $e(t)$, is given by

$$E = \frac{X}{1 + G e^{-sh}}. \tag{4.126}$$

In particular, if the transfer function $G(s) = K/s$, (4.126) reduces to

$$E = \left(\frac{s}{s + K e^{-sh}}\right) X. \tag{4.127}$$

Assuming that the system is quiescent for $t \leqslant 0$, i.e. $x(t) = e(t) = 0$ for $t \leqslant 0$, the delay-differential equation corresponding to (4.127) is

$$\dot{e}(t) + Ke(t - h) = \dot{x}(t). \tag{4.128}$$

For example, for a step input $x(t) = H(t)$ and $X(s) = 1/s$, which means that (4.127) becomes

$$E = \frac{1}{s + K e^{-sh}}, \tag{4.129}$$

which we recognize as the example considered in section 4.2. The corresponding delay-differential equation is

$$\dot{e}(t) + K e(t - h) = \delta(t) \tag{4.130}$$

The effect of the $\delta(t)$ on the right-hand side of this equation is to cause a jump in $e(t)$ at $t = 0$ (exactly as that for $x(t)$). Consequently an alternative formulation is to write

$$\dot{e}(t) + K e(t - h) = 0 \quad \text{for } t > 0 \tag{4.131}$$

together with the conditions

$$e(t) = 0 \quad \text{for } t < 0 \tag{4.132}$$

$$e(0) = 1 \tag{4.133}$$

The latter is termed the initial condition and the above demonstrates how such conditions may be interpreted as inputs to the system.

In the above it was assumed that there was no stored function in the delay, i.e. that there was no output from the delay prior to $t = h$. Consequently the output $y(t)$ is zero for $0 < t < h$. Let us now suppose that this is not the case and there is a function $\phi(t)$ stored in the delay so that the output from the delay and consequently the output from the system is

$$y(t) = \phi(t) \quad \text{for } 0 < t < h. \tag{4.134}$$

From Fig. 4.3 we see that this is equivalent to an input

$$x(t) = -\phi(t)\{H(t) - H(t - h)\} \tag{4.135}$$

into the same system but with no stored function. (4.128) would then apply and we would have

$$\dot{e}(t) + K e(t - h) = -\dot{\phi}(t)\{H(t) - H(t - h)\} - \phi(0)\delta(t) + \phi(h)\delta(t - h). \tag{4.136}$$

Equivalently, since

$$\mathscr{L}\{x(t)\} = -\int_0^h \phi(t) e^{-st} \, dt$$

$$\equiv -\Phi(s), \tag{4.137}$$

say, we have that

$$sE(s) + K e^{-sh} E(s) = -s\Phi(s) \tag{4.138}$$

and so

$$E(s) = \frac{-s\Phi(s)}{s + K e^{-sh}}. \tag{4.139}$$

This demonstrates that any function stored in the delay may be treated as an input but to the same system without a stored function. Thus for the above system it has been demonstrated that problems involving either initial conditions or stored functions may be reformulated as problems involving inputs. Although attention has been restricted to one particular system in this section, this general result is applicable to any system and consequently it suffices to consider the effect of different inputs on the system. The method of evaluation given in this chapter assumed that the transform of error $E(s)$ was of the general form given in (4.9). Within the context of general inputs, is this a reasonable assumption to make and, if $E(s)$ is not of this form, is it possible to evaluate cost functionals for other forms? To attempt to answer such questions we now consider certain examples. Again for simplicity we restrict attention to the system described in Fig. 4.3 with $G(s) = K/s$ and so (4.127) again applies.

Example 4.1 $x(t) = H(t)$

This is, of course, the unit step input mentioned earlier which gives rise to $X(s) = s^{-1}$ and $E(s) = (s + K e^{-sh})^{-1}$. This we know to be of suitable form and the resulting cost functional J is from (4.8a)

$$J = \frac{\cos Kh}{2K(1 - \sin Kh)}. \tag{4.140}$$

Example 4.2 $x(t) = H(t) - H(t - h)$

This gives rise to $X(s) = (1 - e^{-sh})/s$ and $E(s) = (1 - e^{-sh})/(s + K e^{-sh})$. This is the standard form (4.9) with $A(s) = s$, $B(s) = 1$, $C(s) = K$, $D(s) = -1$. It then follows from (4.25) that

$$J = \operatorname*{res}_{s=iK} \left\{ \frac{2(s + K)(1 - e^{-sh})}{(s^2 + K^2)(s + K e^{-sh})} \right\}$$

$$= \frac{\cos Kh}{K(1 - \sin Kh)} - \frac{1}{K} \tag{4.141}$$

Example 4.3 $x(t) = e^{-pt} H(t) \ (p > 0)$

It follows that $X(s) = (s - p)^{-1}$ and

$$E(s) = \frac{s}{(s + p)(s + K e^{-sh})}.$$

Again this is of the standard form but this time we have

$$A(s) = s(s + p), \quad B(s) = s, \quad C(s) = K(s + p), \quad D(s) + 0.$$

It then follows from (4.25) that

$$J = \sum_{s=p,iK} \text{res} \left\{ \frac{s^2(s - Ke^{-sh})}{(s^2 - p^2)(s^2 + K^2)(s + Ke^{-sh})} \right\}$$

$$= \frac{1}{2(K^2 + p^2)} \left\{ \frac{p(p - Ke^{-ph})}{(p + Ke^{-ph})} + \frac{K \cos Kh}{(1 - \sin Kh)} \right\}. \tag{4.142}$$

Example 4.4 $x(t) = e^{-pt}(H(t) - H(t - h))$

Thus

$$X(s) = \frac{1}{s + p} - \frac{e^{-(s+p)h}}{s + p}$$

and so

$$E(s) = \frac{(1 - Le^{-sh})s}{(s + p)(s + Ke^{-sh})}, \quad L \equiv e^{-ph}.$$

This is also of the standard form with

$$A(s) = s(s + p), \quad B(s) = s, \quad C(s) = K(s + p), \quad D(s) = -Ls.$$

Again using (4.25)

$$J = \sum_{s=p,iK} \text{res} \left\{ \frac{(1 + L^2)s^2 + 2KLs^2 - [2Ls^3 + K(1 + L^2)s^2]e^{-sh}}{(s^2 + K^2)(s^2 - p^2)(s + Ke^{-sh})} \right\}$$

$$= \frac{1}{2(K^2 + p^2)} \left\{ (1 - L^2)p - 2KL + \frac{K(1 + L^2)\cos Kh}{(1 - \sin Kh)} \right\}. \tag{4.143}$$

In all of the above examples $E(s)$ was of the standard form and the resulting cost functional was obtained from the general formula (4.25). It is also worth remarking that the form of the input does not affect the stability of the system. This is determined by the zeros of $(s + Ke^{-sh})$ so we know from Example 2.1 that the only region of stability is given by $0 \leqslant Kh < \pi/2$. However, the fact that the system is stable is not sufficient to guarantee that the cost functional J will exist. For an $E(s)$ of the standard form (4.9) this will be so if the polynomial $A(s)$ dominates, which was the case in all of the above examples.

We now consider the case where $E(s)$ is not necessarily of the standard form. In fact $x(t)$ may not even be known explicitly. For example, in Chapter 1 the output $y(t)$ was written as a convolution integral in terms of the input $x(t)$ and the impulse response $g(t)$ in equation (1.2). In other words, for a given system, $g(t)$ is known and the output $y(t)$ may then be written as a functional of the input $x(t)$ which may be regarded as arbitrary. Is something similar possible for the cost functional J? In other words, is it possible to write J, not as an integral involving the error $e(t)$ but one involving the input $x(t)$?

For simplicity we again consider the same system and so from (4.127) we have that

$$E(s) = \frac{sX(s)}{s + K e^{-sh}} = \frac{1}{s + K e^{-sh}} \int_0^\infty \dot{x}(t) e^{-st} \, dt \qquad (4.144)$$

since $\mathscr{L}\{\dot{x}(t)\} = sX(s)$. It then follows that

$$J = \frac{1}{2\pi i} \int_{-i\infty}^{i\infty} E(s)E(-s) \, ds$$

$$= \frac{1}{2\pi i} \int_{-i\infty}^{i\infty} \left(\frac{1}{s + K e^{-sh}} \int_0^\infty \dot{x}(t) e^{-st} \, dt \right) \left(\frac{1}{-s + K e^{sh}} \int_0^\infty \dot{x}(r) e^{st} \, dr \right) ds.$$

Then, assuming that the order of integration may be changed, we obtain

$$J = \frac{1}{2\pi i} \int_0^\infty dt \, \dot{x}(t) \int_0^\infty dr \, \dot{x}(r) \int_{-i\infty}^{i\infty} \frac{e^{-st} e^{sr}}{(s + K e^{-sh})(-s + K e^{sh})} \, ds$$

or, with

$$Q(u) = \frac{1}{2\pi i} \int_{-i\infty}^{i\infty} \frac{e^{su} \, ds}{(s + K e^{-sh})(-s + K e^{sh})} \qquad (4.145)$$

we have that

$$J = \int_0^\infty dt \, \dot{x}(t) \int_0^\infty dr \, \dot{x}(r) Q(r - t). \qquad (4.146)$$

Thus if the function $Q(u)$ can be determined, (4.146) will give the cost functional J as a functional of the input $x(t)$ and we see that it is similar in form to a double convolution integral. We also note that it is $\dot{x}(t)$ that occurs explicitly in this equation rather than $x(t)$ itself. This is to avoid convergence problems. If we had worked in terms of $x(t)$ then the corresponding integral $Q(u)$ would have contained an additional factor $(-s^2)$ in its integrand and consequently the integral would not have existed.

Our aim is therefore to evaluate $Q(u)$ as defined in (4.145); however, before attempting this we note that, by symmetry, $Q(u) = Q(-u)$ and so it suffices to evaluate it for $u \geqslant 0$. We also note that such an integrand would not arise from the standard form for $E(s)$. So we must investigate to see if our method of evaluation is still applicable or if it requires modification.

The first step is to make use of the fact that $u \geq 0$ and to close the contour on the left since the contribution from the semi-circle at infinity there is zero. Thus

$$Q(u) = \sum_{\text{l.h.p}} \text{res} \left\{ \frac{e^{su}}{(s + K e^{-sh})(-s + K e^{sh})} \right\} \tag{4.147}$$

in which the summation is taken over all the poles in the left half-plane which, since the system is assumed to be stable, are the zeros of $(s + K e^{-sh})$. Moreover at such a zero, $e^{sh} = -K/s$ and so (4.147) gives

$$Q(u) = -\sum_{\text{l.h.p}} \text{res} \left\{ \frac{s\, e^{su}}{(s + K e^{-sh})(s^2 + K^2)} \right\}. \tag{4.148}$$

The normal procedure is now to change from a summation over the zeros of $(s + K e^{-sh})$ to one over the zeros of $(s^2 + K^2)$, which are finite in number. To do this it is necessary that the sum of all the residues be zero and this will be true if and only if the integral around the circle at infinity is zero. However, for the expression given in (4.148) this is not true in its present form since $e^{su} \to \infty$ as $|s| \to \infty$ with $\text{Re}\, s > 0$. It is therefore necessary to obtain a more suitable, but equivalent, expression.

To achieve this we introduce $N = [u/h] + 1$ where the square brackets again denote the integer part. We may then write

$$e^{su} = e^{sNh} e^{-s\tau} \tag{4.149}$$

where $\tau = Nh - u$ and so satisfies $0 < \tau \leq h$. Consequently at a zero of $(s + K e^{-sh})$, $e^{sh} = -K/s$ and so

$$e^{su} = \left(-\frac{K}{s}\right)^N e^{-s\tau} \tag{4.150}$$

and (4.148) may be written

$$Q(u) = -\sum_{\text{l.h.p}} \text{res} \left\{ \frac{(-K)^N e^{-s\tau}}{s^{N-1}(s^2 + K^2)(s + K e^{-sh})} \right\} \tag{4.151}$$

For the expression in this equation the integral round the circle at infinity is indeed zero and so the sum of all the residues is zero. Thus

$$Q(u) = \sum_{s=0,\pm iK} \text{res} \left\{ \frac{(-K)^N e^{-s\tau}}{s^{N-1}(s^2 + K^2)(s + K e^{-sh})} \right\}. \tag{4.152}$$

Now $N \geq 1$ and if $N = 1$ there is no residue at $s = 0$. Otherwise

$$\text{res}_{s=0} \left\{ \frac{(-K)^N e^{-s\tau}}{s^{N-1}(s^2 + K^2)(s + K e^{-sh})} \right\}$$

$$= (N-2)! \left(\frac{\partial}{\partial s}\right)^{N-2} \left\{ \frac{(-K)^N e^{-s\tau}}{(s^2 + K^2)(s + K e^{-sh})} \right\} \bigg|_{s=0}. \tag{4.153}$$

Secondly

Sec. 4.7] **Stored functions, arbitrary inputs and initial conditions** 81

$$\sum_{s=\pm iK} \left\{ \frac{(-K)^N e^{-s\tau}}{s^{N-1}(s^2+K^2)(s+Ke^{-sh})} \right\}$$

$$= (-K)^N \left\{ \frac{e^{-iK\tau}}{2(iK)^N(iK+Ke^{-iKh})} + \frac{e^{iK\tau}}{2(-iK)^N(-iK+Ke^{iKh})} \right\}$$

$$= \frac{\sin\left(\frac{N\pi}{2} - K\tau\right) + \cos\left(\frac{N\pi}{2} + K(h-\tau)\right)}{2K(1-\sin Kh)}. \tag{4.154}$$

Thus

$$Q(u) = (N-2)! \left(\frac{\partial}{\partial s}\right)^{N-2} \left\{ \frac{(-K)^N e^{-s\tau}}{(s^2+K^2)(s+Ke^{-sh})} \right\} \bigg|_{s=0}$$

$$+ \frac{\sin\left(\frac{N\pi}{2} - K\tau\right) + \cos\left(\frac{N\pi}{2} + K(h-\tau)\right)}{2K(1-\sin Kh)} \tag{4.155}$$

for $u > 0$ and, from (4.146),

$$J = \int_0^\infty dt\, \dot{x}(t) \int_0^\infty dr\, \dot{x}(r) Q(|r-t|). \tag{4.156}$$

Thus, in principle, J may be determined for any input.

It follows that within the context of this chapter there are two routes to the evaluation of J. For a given $x(t)$, $\dot{x}(t)$ may be determined and J may then be obtained from (4.156). This requires the evaluation of the two integrals in this equation. Alternatively, the Laplace transforms $X(s)$, and hence $E(s)$, may be determined and, provided the latter is of the standard form, evaluation proceeds as before.

It is worth remembering that (4.156) applies only to systems described by Fig. 4.3 with $G(s) = K/s$, or equivalently by (4.128). For more general systems the analogous expression could also contain higher-order derivatives of $x(t)$.

One particular example worth considering in more detail is that when the input corresponds to that from a function $\phi(t)$ stored in the delay and so from (4.135)

$$x(t) = -\phi(t)\{H(t) - H(t-h)\}. \tag{4.157}$$

Differentiating this we obtain

$$\dot{x}(t) = -\dot{\phi}(t)\{H(t) - H(t-h)\} - \phi(0)\delta(t) + \phi(h)\delta(t-h). \tag{4.158}$$

With this expression (4.156) becomes

$$J = \int_0^h dt\,\dot\phi(t) \int_0^h dr\,\dot\phi(r) Q(|r-t|) + 2\phi(0)\int_0^h \dot\phi(t) Q(t)\,dt$$

$$- 2\phi(h)\int_0^h \dot\phi(t) Q(h-t)\,dt + \{[\phi(0)]^2 + [\phi(h)]^2\}Q(0)$$

$$- 2\phi(0)\phi(h)Q(h). \tag{4.159}$$

Moreover $|r-t| \leqslant h$ and so it suffices to know $Q(u)$ for those values of u satisfying $0 \leqslant u \leqslant h$. Now for $0 \leqslant u < h$, $n = 1$ and hence $\tau = h - u$ and

$$Q(u) = \frac{\cos K\tau - \sin K(h-\tau)}{2K(1 - \sin Kh)}$$

$$= \frac{\cos K(h-u) - \sin Ku}{2K(1 - \sin Kh)}. \tag{4.160}$$

In particular,

$$Q(0) = \frac{\cos Kh}{2K(1 - \sin Kh)}. \tag{4.161}$$

Secondly for $u = h$, $N = 2$ and hence $\tau = h$ and

$$Q(h) = \frac{1}{K} + \frac{\sin K\tau - \cos K(h-\tau)}{2K(1 - \sin Kh)} = \frac{1}{2K}. \tag{4.162}$$

We therefore conclude that

$$J = \int_0^h dt\,\dot\phi(t) \int_0^h dr\,\dot\phi(r) \left\{\frac{\cos K(h-|r-t|) - \sin K|r-t|}{2K(1 - \sin Kh)}\right\}$$

$$+ 2\phi(0)\int_0^h \dot\phi(t)\left\{\frac{\cos K(h-t) - \sin Kt}{2K(1 - \sin Kh)}\right\}dt$$

$$- 2\phi(h)\int_0^h \dot\phi(t)\left\{\frac{\cos Kt - \sin K(h-t)}{2K(1 - \sin Kh)}\right\}dt$$

$$+ \{[\phi(0)]^2 + [\phi(h)]^2\}\left\{\frac{\cos Kh}{2K(1 - \sin Kh)}\right\} - \frac{1}{K}\phi(0)\phi(h). \tag{4.163}$$

Although the whole of this section has been restricted to the system given in Fig. 4.3 with $G(s) = K/s$ the ideas and methods presented are widely applicable. The sole reason for this restriction was to avoid as much algebraic complexity as possible.

5

The Lyapunov method

5.1 INTRODUCTION

In this chapter we shall use the description of the control system in the form of a set of delay-differential equations

$$\dot{x}(t) = Ax(t) + Bx(t - h) + Cu(t), \tag{5.1}$$

a generalization of the delay-free equation (1.7). In (5.1), $x(t)$ is an n-dimensional vector, $u(t)$ is an m-dimensional input vector, h is a positive time delay, and A, B and C are constant matrices of compatible dimensions. The vector equation (5.1) describes the dynamics of the system; the measurable or observable quantities are given by the output equation associated with (5.1)

$$y(t) = Dx(t) \tag{5.2}$$

where $y(t)$ is the output vector of dimension p and D is a constant $p \times n$ matrix. Equations (5.1) and (5.2) fully characterize the open-loop control system; a typical closed-loop structure is shown in Fig. 5.1. w denotes a constant reference input. In this case $p = m$ and $u(t) = w - y(t)$.

A convenient and frequently used means of estimating the system performance is the quadratic integral criterion as introduced in section 1.6. This can be in the form

$$J_1 = \int_0^\infty y(t)^T R y(t) \, dt \tag{5.3}$$

or in the form of integral square error

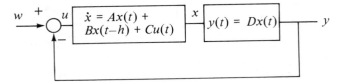

Fig. 5.1. Closed-loop control system with delay.

$$J_2 = \int_0^\infty u(t)^T R u(t)\, dt, \text{ where } u = w - y. \tag{5.4}$$

R denotes a symmetric and positive semidefinite matrix of weighting factors.
For our purposes we shall reduce both criteria (5.3) and (5.4) to the form

$$J = \int_0^\infty x(t)^T V x(t)\, dt \tag{5.5}$$

where V is a positive semidefinite and symmetric matrix. The transformation of (5.3) into (5.5) is straightforward. In order to transform (5.4) we may extend the set of equations (5.1) by adding to it

$$\dot{u}(t) = \dot{w} - \dot{y}(t)$$
$$= -DAx(t) - DBx(t - h) - DCu(t)$$
$$u(0) = w - y(0). \tag{5.6}$$

If we now define a new $(n + m)$-dimensional vector z and an $(n + m) \times (n + m)$-dimensional matrix V,

$$z = \begin{pmatrix} x \\ u \end{pmatrix}, \quad V = \begin{pmatrix} 0 & 0 \\ 0 & R \end{pmatrix}, \tag{5.7}$$

we may rewrite (5.4) in the form

$$J_2 = \int_0^\infty z(t)^T V z(t)\, dt. \tag{5.8}$$

Let us note that it is not always necessary to increase the dimension of the problem by m. In order to prove this, consider the extreme case where no increase of dimension occurs. If there exists a constant vector a such that

$$Da = w \tag{5.9}$$

and

$$(A + B)a = 0 \tag{5.10}$$

then the simple substitution

$$x(t) = z(t) + a \tag{5.11}$$

yields the system equation in the form

$$\dot{z}(t) = (A - CD)z(t) + Bz(t - h). \tag{5.12}$$

For the performance index (5.4) we obtain

$$J_2 = \int_0^\infty z(t)^T D^T R D z(t) \, dt. \tag{5.13}$$

The problem of evaluation of integral square error for many practical control systems with delays has a form which is apparently different from the one described above. And yet, in some cases, an appropriate transformation of variables yields a strictly quadratic performance index (5.5) and the system equation in a homogeneous form.

We shall now consider two simple examples.

Example 5.1
The regulation system is shown in Fig. 5.2. The controlled plant is described by the transfer function

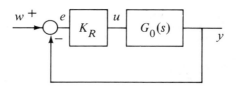

Fig. 5.2. Regulation system of Example 5.1.

$$G_0(s) = \frac{K_0 \exp(-sh)}{s}, \quad K_0 > 0 \tag{5.14}$$

and the regulator is of P-type with gain $K_R > 0$. The reference input w is constant, the system is in the state of equilibrium until $t = 0$, and at $t = 0$ the value of y is changed from y_∞ to some y_0. The quadratic criterion is

$$J = \int_0^\infty [y(t) - y_\infty]^2 \, dt. \tag{5.15}$$

In the state of equilibrium $y(t) = y_\infty = w$. The system dynamics is described by the equation

$$\dot{y}(t) = K[w - y(t - h)], \quad t \geqslant 0, \quad K = K_0 K_R. \tag{5.16}$$

Denoting $x(t) = y(t) - w$ we come to the following problem. Compute

$$J = \int_0^\infty x(t)^2 \, dt \tag{5.17}$$

where x is the solution of the initial value problem

$$\dot{x}(t) = -Kx(t-h), \quad t \geq 0$$
$$x(0) = y_0 - w, \quad x(t) = 0, \quad t < 0. \tag{5.18}$$

We remark that this system is equivalent to that introduced in section 2.2 and discussed in some detail in Chapter 2. It also gives rise to the simple example considered in sections 3.2 and 4.2.

Example 5.2

Let us consider the regulation system shown in Fig. 5.3. The transfer function of the controlled plant is

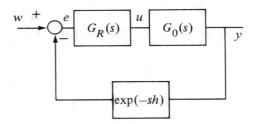

Fig. 5.3. Regulation system of Example 5.2.

$$G_0(s) = \frac{K_0}{T_0 s + q}, \quad T_0, K_0 > 0, \quad q = 0 \text{ or } 1. \tag{5.19}$$

The regulator is of I or PI-type,

$$G_R(s) = K_R \left(p_0 + \frac{1}{T_i s} \right), \quad T_i, K_R > 0, \quad p_0 = 0 \text{ or } 1. \tag{5.20}$$

The delay is in the feedback loop. We assume that the reference input w is constant and the system is in the state of equilibrium until $t = 0$, and at $t = 0$ a delta-type disturbance changes the value of y from y_∞ to some y_0. The integral performance criterion is of the form

$$J = \int_0^\infty [y(t) - y_\infty]^2 \, dt. \tag{5.21}$$

Denoting $e(t) = w - y(t - h)$ we obtain the following equation for the system dynamics:

$$T_0 \dot{y}(t) + qy(t) = p_0 K e(t) + \frac{K}{T_i} \int_0^t e(s) \, ds + C, \quad t > 0 \quad (5.22)$$

where C is a constant and $K = K_0 K_R$. From this equation it is evident that $y_\infty = w$ and $C = qw$. Define

$$x_2(t) = y(t) - w, \quad x_1(t) = \int_0^t x_2(s) \, ds \quad (5.23)$$

and

$$a = \frac{q}{T_0}, \quad b = \frac{K}{T_i T_0}, \quad c = \frac{p_0 K}{T_0}. \quad (5.24)$$

The equations of the regulation system then take the form

$$\dot{x}_1(t) = x_2(t)$$
$$\dot{x}_2(t) = -ax_2(t) - bx_1(t - h) - cx_2(t - h) \quad (5.25)$$

and the performance index is

$$J = \int_0^\infty x_2(t)^2 \, dt. \quad (5.26)$$

The study of Lyapunov functionals and Lyapunov systems of equations for time-delay control systems has received considerable attention for about twenty years. Although most of the papers were devoted to study of stability, and not to the evaluation of integral performance criteria, many of the results can be directly applied to our purposes. Repin (1965) was the first to apply the Lyapunov method to time-delays systems and obtain analytic expressions for integrals of type (5.5). However, his results were correct only for the one-dimensional case. The Lyapunov system of equations for system (5.1) and functional (5.5) was constructed by many authors, e.g. Datko (1972), Delfour, McCalla and Mitter (1975), Infante and Castelan (1978), Castelan and Infante (1979), Grabowski (1983), often as a particular case of more general problems (many lumped and distributed delays—Delfour, McCalla and Mitter (1975); neutral systems—Castelan and Infante (1979), Grabowski (1983); delay in the integrand—Grabowski (1983)). Delfour, McCalla and Mitter (1975) give the Lyapunov system in a form different from the one used in this book, in which the 'counterflow' functional–differential equation is replaced by one ordinary and one partial differential equation. However, the 'counterflow' equation may be immediately deduced from their results. For the simplest problem of evaluating integral (5.5) on trajectories of (5.1) with $u = 0$, we use the Lyapunov system of equations in the form proposed in Grabowski (1983) for a more general case. This in turn is a development of the results of Infante and Castelan (1978), and Castelan and Infante (1979). Of particular significance is the analysis in Castelan and Infante (1977) of the 'counterflow' equation appearing in the Lyapunov set of equations. We follow the general ideas given there but use a different transformation of the 'counterflow' equation into a set of ordinary linear differential equations which allows us to reduce the dimensionality of the problem and obtain explicit dependence of the characteristic polynomial on λ^2. The evaluation of performance criteria with weighting functions $t^k \exp(at)$ is based on differentiation under the integral sign, an old trick suggested to us by J. Sularz, one of our doctoral students. This idea enabled us to derive a recursive system of Lyapunov-type equations. The results concerning the functional with squared derivatives, and the solution of the 'counterflow' equation for the case of many commensurate delays with the use of the step method, are simple new developments. Most of the results concerning neutral systems with one delay may be found in Castelan and Infante (1979) and Grabowski (1983a), although we obtain them in a more direct way, not via the construction of Lyapunov functionals.

The Lyapunov approach for delay-free systems was considered in detail in section 1.6 and the main results may be summarized briefly as follows. For a system described by the vector equation

$$\dot{x}(t) = Ax(t) \tag{5.27}$$

we may write

$$x(t) = \Phi(t)x(0) \tag{5.28}$$

where

$$\Phi(t) = \exp(At) \tag{5.29}$$

denotes the fundamental matrix and satisfies

$$\Phi A = A\Phi. \tag{5.30}$$

The quadratic performance criterion (5.5) may be written

$$J = x(0)^T M x(0) \tag{5.31}$$

where

$$M = \int_0^\infty \Phi(t) V \Phi(t) \, dt \tag{5.32}$$

is a symmetric and positive semidefinite matrix which may be shown to satisfy the Lyapunov equation

$$MA + A^T M + V = 0. \tag{5.33}$$

5.2. CASE OF A SINGLE DELAY AND POINTWISE INITIAL CONDITIONS

Let us now consider the Lyapunov approach for a simple problem with one delay. We shall calculate the performance index (5.5) on trajectories of the system

$$\dot{x}(t) = Ax(t) + Bx(t - h). \tag{5.34}$$

We assume that $x(0)$ is a given arbitrary vector in $\mathbb{R}^n$ and

$$x(t) = 0, \quad t < 0. \tag{5.35}$$

In order to indicate the dependence of the performance criterion J on the initial condition $x(0)$ we shall write $J(x(0))$.

The fundamental matrix solution Φ of (5.34) is defined by

$$\dot{\Phi}(t) = A\Phi(t) + B\Phi(t - h)$$

$$\Phi(0) = I, \quad \Phi(t) = 0 \quad \text{for } t < 0. \tag{5.36}$$

The solution of (5.34) can be written in the form

$$x(t) = \Phi(t)x(0). \tag{5.37}$$

The substitution of (5.37) into (5.5) gives

$$J(x(0)) = x(0)^T M x(0) \tag{5.38}$$

where, as before,

$$M = \int_0^\infty \Phi(t)^T V \Phi(t) \, dt$$

$$= \int_0^\infty \Phi(t - s)^T V \Phi(t - s) \, dt, \quad \forall s \geq 0. \tag{5.39}$$

We differentiate this with respect to s and substitute $s = 0$ which gives

$$0 = \int_s^\infty [A\Phi(t) + B\Phi(t - h)]^T V \Phi(t) \, dt$$

$$+ \int_0^\infty \Phi(t)^T V [A\Phi(t) + B\Phi(t - h)] \, dt + V. \tag{5.40}$$

Now we use the following identity:

$$A\Phi(t) + B\Phi(t - h) = \Phi(t)A + \Phi(t - h)B \tag{5.41}$$

which can be easily proved by means of Laplace transform techniques. Hence we obtain

$$MA + A^T M + V + L(h)^T + L(h) = 0 \tag{5.42}$$

where

$$L(h) = \int_0^\infty \Phi(t)^T V \Phi(t-h) B \, dt$$

$$= \int_0^\infty \Phi(t+h)^T V \Phi(t) B \, dt. \tag{5.43}$$

(5.42) is the analogue of the delay-free equation (5.33) and contains the additional terms in $L(h)$. Consequently to determine M, further equations are required. These are obtained by considering the more general function

$$L(s) = \int_0^\infty \Phi(t+s)^T V \Phi(t) B \, dt, \quad s \in [0, h]. \tag{5.44}$$

To obtain a differential equation for L, we differentiate both sides of equality (5.44) with respect to s which yields

$$\dot{L}(s) = A^T L(s) + L(h-s)^T B.$$

Hence we have obtained the so-called Lyapunov set of equations for determining M and hence $J(x(0))$:

$$MA + A^T M + V + L(h)^T + L(h) = 0 \tag{5.45}$$

$$\dot{L}(s) = A^T L(s) + L(h-s)^T B, \quad s \in [0, h]. \tag{5.46}$$

$$L(0) = MB. \tag{5.47}$$

Before carrying on the general analysis, we shall solve this set of equations for the simple particular case of Example 5.1 from section 5.1. We have there $n = 1$, $A = 0$, $B = -K$ and $V = 1$, and so equations (5.45)–(5.47) take the form

$$1 + 2L(h) = 0$$

$$\dot{L}(s) = -KL(h-s), \quad s \in [0, h]$$

$$L(0) = -KM. \tag{5.48}$$

The functional–differential equation which appears here may be easily solved using the substitution

$$Z(s) = L(h-s). \tag{5.49}$$

We thus obtain a set of ordinary linear differential equations

$$\dot{L}(s) = -KZ(s)$$

$$\dot{Z}(s) = KL(s) \tag{5.50}$$

with the boundary conditions

$$L(h) = Z(0) = -\tfrac{1}{2}, \quad L(0) = Z(h) = -KM. \tag{5.51}$$

The general solution of (5.50) is

$$\begin{pmatrix} L(s) \\ Z(s) \end{pmatrix} = \begin{pmatrix} \cos Ks & -\sin Ks \\ \sin Ks & \cos Ks \end{pmatrix} \begin{pmatrix} c_1 \\ c_2 \end{pmatrix}. \tag{5.52}$$

Choose the constants c_1 and c_2 so that $Z(s) = L(h - s)$, $\forall s$, that is,

$$c_1 \sin Ks + c_2 \cos Ks = c_1 \cos K(h - s) - c_2 \sin K(h - s), \forall s.$$

Hence we obtain one linear relationship between c_1 and c_2,

$$c_1 = c_2 \frac{\cos Kh}{1 - \sin Kh}.$$

Two parameters, c_2 and M, are still to be determined. These we calculate from the boundary conditions. Since $Z(0) = -\tfrac{1}{2}$, we obtain $c_2 = -\tfrac{1}{2}$, and $L(0) = -KM$ yields $c_1 = -KM$. Hence finally

$$M = \frac{\cos Kh}{2K(1 - \sin Kh)} = \frac{1 + \sin Kh}{2K \cos Kh}, \tag{5.53}$$

$$L(s) = \frac{\sin Ks - \cos K(s - h)}{2(1 - \sin Kh)} = \frac{\sin K(s - h) - \cos Ks}{2 \cos Kh}. \tag{5.54}$$

As was to be expected the result given in (5.53) is exactly that given in (3.22) of section 3.2 and (4.8) of section 4.2.

Let us return to the general case. The set of matrix equations (5.45)–(5.47) is the generalization of equation (5.33) in the delay-free case. Notice that if $h = 0$, the system reduces to (5.33) with A replaced by $A + B$. If we put $B = 0$, we have $L = 0$ and once more obtain (5.33).

The system (5.45)–(5.47) will have a solution with positive semidefinite and symmetric M, if integral (5.5) exists for every $x(0)$ in $\mathbb{R}^n$. A sufficient condition for this is that system (5.34) is asymptotically stable, i.e. all roots of the characteristic equation

$$\det [\lambda I - A - B \exp(-\lambda h)] = 0 \tag{5.55}$$

have negative real parts. Moreover, if system (5.34) is asymptoticaly stable, then system (5.45)–(5.47) has exactly one solution such that $M = M^T \geqslant 0$. Let us also note that asymptotic stability of (5.34) implies that equation (5.46) with initial condition (5.47) has a unique solution L for every symmetric, positive semidefinite M.

5.3 SOLUTION OF THE LYAPUNOV SYSTEM OF EQUATIONS

The starting point in the analysis of the Lyapunov set of equations (5.45)–(5.47) is a transformation of the functional–differential equation (5.46), called the 'counterflow' equation, with initial condition (5.47) into a set of ordinary linear equations with constant coefficients. We define

$$U(s) = \tfrac{1}{2}[L(s) + L(h - s)]$$
$$W(s) = \tfrac{1}{2}[L(s) - L(h - s)] \tag{5.56}$$

whence

$$L(s) = U(s) + W(s)$$
$$L(h - s) = U(s) - W(s) \tag{5.57}$$

and rewrite (5.46) with condition (5.47) in the form of two linear matrix differential equations

$$\dot{U}(s) = A^{\mathrm{T}}W(s) - W(s)^{\mathrm{T}}B, \quad s \in [0, h]$$
$$\dot{W}(s) = A^{\mathrm{T}}U(s) + U(s)^{\mathrm{T}}B \tag{5.58}$$

with the condition

$$U(0) + W(0) = U(h) - W(h) = MB. \tag{5.59}$$

U is symmetric and W is antisymmetric with respect to $s = \tfrac{1}{2}h$, which means that

$$U(s) = U(h - s), \quad W(s) = -W(h - s), \quad s \in [0, h]. \tag{5.60}$$

This initial value problem is equivalent to (5.46), (5.47). System (5.58) can easily be transformed into a set of ordinary linear differential equations where the unknowns are elements of the matrices $U(s)$ and $W(s)$. To this end we define

$$\tilde{L} = (L^{(1)^{\mathrm{T}}}, L^{(2)^{\mathrm{T}}}, \ldots, L^{(n)^{\mathrm{T}}})$$
$$\tilde{U} = (U^{(1)^{\mathrm{T}}}, U^{(2)^{\mathrm{T}}}, \ldots, U^{(n)^{\mathrm{T}}})$$
$$\tilde{W} = (W^{(1)^{\mathrm{T}}}, W^{(2)^{\mathrm{T}}}, \ldots, W^{(n)^{\mathrm{T}}}). \tag{5.61}$$

Here, $L^{(i)}$, $U^{(i)}$ and $W^{(i)}$ denote the ith columns of L, U and W, respectively. We then obtain from (5.58)

$$(\dot{\tilde{U}}(s), \dot{\tilde{W}}(s)) = (\tilde{U}(s), \tilde{W}(s))\, N \tag{5.62}$$

where

$$N = \begin{pmatrix} 0 & \operatorname{diag} A + \beta \\ \operatorname{diag} A - \beta & 0 \end{pmatrix} \begin{matrix} \}n^2 \\ \}n^2 \end{matrix} \tag{5.63}$$
$$\underbrace{\phantom{\operatorname{diag} A - \beta}}_{n^2} \underbrace{\phantom{\operatorname{diag} A + \beta}}_{n^2}$$

$$\beta = (\operatorname{diag} B^{(1)}, \operatorname{diag} B^{(2)}, \ldots, \operatorname{diag} B^{(n)}). \tag{5.64}$$

$B^{(i)}$ is the ith column of B, and diag Y for any matrix Y denotes a block-diagonal matrix with blocks Y on the diagonal and zeros elsewhere.

The standard theory applied to system (5.62) gives the following results. The solution is expressed by the formula

$$(\tilde{U}(s), \tilde{W}(s)) = (\tilde{U}(s_0), \tilde{W}(s_0)) \exp[N(s - s_0)] \tag{5.65}$$

for any $s_0 \in [0, h]$. If we denote

Sec. 5.3] Solution of the Lyapunov system of equations

$$P = \text{diag } A + \beta, \quad Q = \text{diag } A - \beta, \quad z = s - \frac{h}{2} \tag{5.66}$$

then

$$\exp(Nz) = \sum_{i=0}^{\infty} \begin{pmatrix} \dfrac{(PQ)^i z^{2i}}{(2i)!} & P\dfrac{(QP)^i z^{2i+1}}{(2i+1)!} \\ Q\dfrac{(PQ)^i z^{2i+1}}{(2i+1)!} & \dfrac{(QP)^i z^{2i}}{(2i)!} \end{pmatrix} \tag{5.67}$$

It can be readily seen that both the symmetry conditions (5.60) are equivalent to the requirement

$$\tilde{W}\left(\frac{h}{2}\right) = 0. \tag{5.68}$$

After using this condition, the solution of (5.62) is determined uniquely up to an unknown $n \times n$ matrix of coefficients, equal to $U(h/2)$. This solution can be written in an explicit form

$$\tilde{U}(s) = \tilde{U}\left(\frac{h}{2}\right) \sum_{i=0}^{\infty} \frac{(PQ)^i z^{2i}}{(2i)!}$$

$$\tilde{W}(s) = \tilde{U}\left(\frac{h}{2}\right) \sum_{i=0}^{\infty} \frac{(PQ)^i z^{2i+1}}{(2i+1)!} \cdot P. \tag{5.69}$$

Hence

$$\tilde{L}(s) = \tilde{U}\left(\frac{h}{2}\right) \sum_{i=0}^{\infty} \frac{(PQ)^i z^{2i}}{(2i)!} \left(I + \frac{Pz}{2i+1}\right). \tag{5.70}$$

This series representation of the solution may be useful in practice, in particular, if the delay h is small. The convergence of this series is then very fast in the whole interval $[0, h]$.

We shall now give finite explicit formulae for the solution, based on the technique of characteristic roots and characteristic vectors. The characteristic equation is

$$\det(N - \lambda I) = 0. \tag{5.71}$$

We shall denote by I any unit matrix; its dimension will always be evident from the context. In order to simplify the characteristic equation, notice that

$$\det N = \det[-(\text{diag } A - \beta)(\text{diag } A + \beta)] \tag{5.72}$$

and for $\lambda \neq 0$,

$$\det(N - \lambda I) = \det[Q(\lambda)(N - \lambda I)] \tag{5.73}$$

with

$$Q(\lambda) = \begin{pmatrix} I & 0 \\ \lambda^{-1}(\text{diag } A - \beta) & I \end{pmatrix}. \tag{5.74}$$

Finally, the characteristic equation takes the form

$$\det[\lambda^2 I - (\text{diag } A - \beta)(\text{diag } A + \beta)] = 0. \tag{5.75}$$

The left-hand side of the characteristic equation, called the characteristic polynomial, is a polynomial of degree n^2 in the variable $\rho = \lambda^2$. Therefore, if λ_i is a root of the characteristic polynomial (or a characteristic root), then $-\lambda_i$ is also a characteristic root. Let us note that in the Jordan canonical form of N the blocks on the diagonal which correspond to the roots λ_i and $-\lambda_i$ are identical in number and structure. With each characteristic root λ_i we associate characteristic row vectors X_i of dimension $2n^2$, determined by the equation

$$X_i(N - \lambda_i I) = 0, \quad X_i \neq 0. \tag{5.76}$$

There are $2n^2 - \text{rank}(N - \lambda_i I)$ linearly independent characteristic vectors, corresponding to the characteristic root λ_i. Denote

$$X_i = (\tilde{Y}_i, \tilde{Z}_i) \tag{5.77}$$

where $\tilde{Y}_i$ and $\tilde{Z}_i$ are n^2-dimensional row vectors. Equation (5.76) is equivalent to

$$\lambda_i \tilde{Y}_i = \tilde{Z}_i(\text{diag } A - \beta)$$
$$\lambda_i \tilde{Z}_i = \tilde{Y}_i(\text{diag } A + \beta). \tag{5.78}$$

In the case of $\lambda_i \neq 0$ we have

$$\tilde{Y}_i = \lambda_i^{-1} \tilde{Z}_i(\text{diag } A - \beta) \tag{5.79}$$

and

$$\tilde{Z}_i[\lambda_i^2 I - (\text{diag } A - \beta)(\text{diag } A + \beta)] = 0. \tag{5.80}$$

We recognize here the matrix whose determinant is the characteristic polynomial (see (5.75), with $\lambda = \lambda_i$).

Let us assume that the characteristic equation (5.75) has only single roots λ_i. The general solution of system (5.62) then has the form

$$(\tilde{U}(s), \tilde{W}(s)) = \sum_i [a_i(\tilde{Y}_i, \tilde{Z}_i)\exp(\lambda_i s) + b_i(-\tilde{Y}_i, \tilde{Z}_i)\exp(\lambda_i(h - s))] \tag{5.81}$$

where the summation is over all pairs of characteristic roots $(\lambda_i, -\lambda_i)$, a_i and b_i are arbitrary constants, $\tilde{Z}_i$ is a non-zero solution of (5.80) and $\tilde{Y}_i$ is determined from (5.79). Let us now define $n \times n$ matrices Y_i and Z_i,

$$\tilde{Y}_i = (Y_i^{(1)^T}, Y_i^{(2)^T}, \ldots, Y_i^{(n)^T}), \quad \tilde{Z}_i = (Z_i^{(1)^T}, Z_i^{(2)^T}, \ldots, Z_i^{(n)^T}) \tag{5.82}$$

where $Y_i^{(j)}$ and $Z_i^{(j)}$ are the jth columns of Y_i and Z_i, respectively. If we use one of the symmetry conditions (5.60) (the other is redundant), we can write

$$U(s) = \sum_i a_i [\exp(\lambda_i s) + \exp(\lambda_i(h-s))] Y_i$$

$$= \sum_i a_i' \cosh\left[\lambda_i\left(s - \frac{h}{2}\right)\right] Y_i \qquad (5.83)$$

$$W(s) = \sum_i a_i [\exp(\lambda_i s) - \exp(\lambda_i(h-s))] Z_i$$

$$= \sum_i a_i' \sinh\left[\lambda_i\left(s - \frac{h}{2}\right)\right] Z_i$$

where $a_i' = 2a_i \exp(\lambda_i h/2)$. Hence, by virtue of (5.57)

$$L(s) = \sum_i a_i [(Y_i + Z_i)\exp(\lambda_i s) + (Y_i - Z_i)\exp(\lambda_i(h-s))]$$

$$= \sum_i a_i' \left[Y_i \cosh\left(\lambda_i\left(s - \frac{h}{2}\right)\right) + Z_i \sinh\left(\lambda_i\left(s - \frac{h}{2}\right)\right) \right]. \qquad (5.84)$$

The summation is over all pairs of roots $(\lambda_i, -\lambda_i)$ of equation (5.75).

It is easy to check that the characteristic matrices Y_i and Z_i satisfy the following set of matrix equations:

$$\lambda_i Y_i = A^T Z_i - Z_i^T B$$
$$\lambda_i Z_i = A^T Y_i + Y_i^T B. \qquad (5.85)$$

Let us now discuss the case where the characteristic equation (5.75) has multiple roots λ_i. Notice that if $\lambda_i = 0$ is a characteristic root, it must have an even multiplicity. Assume that p_i blocks of the form

$$\begin{pmatrix} \lambda_i & & & & 0 \\ 1 & \lambda_i & & & \\ & 1 & \ddots & & \\ & & \ddots & \lambda_i & \\ 0 & & & 1 & \lambda_i \end{pmatrix} \qquad (5.86)$$

occur in the Jordan canonical form of N and the dimensions of these blocks are $r_i^{(1)}$, $r_i^{(2)}, \ldots, r_i^{(p_i)}$. With each such block of dimension $r_i^{(j)}$ we associate generalized characteristic vectors $X_i^{(1,j)}, X_i^{(2,j)}, \ldots, X_i^{(q,j)}$, $q = r_i^{(j)}$. The first-order vectors $X_i^{(1,j)}$ are simply the characteristic vectors discussed above. The higher-order generalized characteristic vectors satisfy the equations

$$X_i^{(k,j)}(N - \lambda_i I) = X_i^{(k-1,j)}, \quad k = 2, \ldots, r_i^{(j)}, \qquad (5.87)$$

or, equivalently,

$$X_i^{(k,j)}(N - \lambda_i I)^k = 0 \quad \text{and} \quad X_i^{(k,j)}(N - \lambda_i I)^{k-1} \neq 0, \quad k = 1, 2, \ldots, r_i^{(j)}. \qquad (5.88)$$

If we decompose the vectors $X_i^{(k,j)}$ in the same way as we did for X_i (5.77),

$$X_i^{(k,j)} = (\tilde{Y}_i^{(k,j)}, \tilde{Z}_i^{(k,j)}) \tag{5.89}$$

where $\tilde{Y}_i^{(k,j)}$ and $\tilde{Z}_i^{(k,j)}$ are row n^2-vectors, we obtain from (5.87)

$$-\lambda_i \tilde{Y}_i^{(k,j)} + \tilde{Z}_i^{(k,j)}(\text{diag } A - \beta) = \tilde{Y}_i^{(k-1,j)}$$

$$-\lambda_i \tilde{Z}_i^{(k,j)} + \tilde{Y}_i^{(k,j)}(\text{diag } A + \beta) = \tilde{Z}_i^{(k-1,j)}, \quad k = 2, \ldots, r_i^{(j)}. \tag{5.90}$$

By analogy to the case of single characteristic roots, we introduce generalized characteristic matrices $Y_i^{(k,j)}$, $Z_i^{(k,j)}$ connected with $X_i^{(k,j)}$ in the same fashion as the matrices Y_i and Z_i with X_i (5.82). It is easy to check that they satisfy the following set of matrix equations

$$A^T Z_i^{(k,j)} - Z_i^{(k,j)^T} B - \lambda_i Y_i^{(k,j)} = Y_i^{(k-1,j)}$$

$$A^T Y_i^{(k,j)} + Y_i^{(k,j)^T} B - \lambda_i Z_i^{(k,j)} = Z_i^{(k-1,j)}, \quad k = 2, \ldots, r_i^{(j)}. \tag{5.91}$$

The total number of linearly independent generalized characteristic vectors is $2n^2$, the same as the number of linearly independent pairs $(Z_i^{(k,j)}, Y_i^{(k,j)})$ of generalized characteristic matrices.

Let us also introduce matrices

$$F_i^{(k,j)} = Y_i^{(k,j)} + Z_i^{(k,j)}$$

$$H_i^{(k,j)} = Y_i^{(k,j)} - Z_i^{(k,j)}, \quad k = 1, 2, \ldots, r_i^{(j)}. \tag{5.92}$$

In this notation, system (5.91) takes a simpler form

$$(A^T - \lambda_i I)F_i^{(k,j)} + H_i^{(k,j)^T} B = F_i^{(k-1,j)}$$

$$-F_i^{(k,j)^T} B - (A^T + \lambda_i I)H_i^{(k,j)} = H_i^{(k-1,j)}, \quad k = 2, \ldots, r_i^{(j)}. \tag{5.93}$$

We shall now determine the relationship between the generalized characteristic vectors corresponding to λ_i and $-\lambda_i$. We already know that if $(\tilde{Y}_i^{(1,j)}, \tilde{Z}_i^{(1,j)})$ is a characteristic vector corresponding to λ_i, then $(-\tilde{Y}_i^{(1,j)}, \tilde{Z}_i^{(1,j)})$ is a characteristic vector corresponding to $-\lambda_i$. Applying an inductive argument to system (5.90) we obtain that if $(\tilde{Y}_i^{(k,j)}, \tilde{Z}_i^{(k,j)})$ is a generalized characteristic vector of order k, corresponding to λ_i, then $(-1)^k(\tilde{Y}_i^{(k,j)}, -\tilde{Z}_i^{(k,j)})$ is a generalized characteristic vector of the same order, corresponding to $-\lambda_i$. This is also true for zero characteristic roots, which means that if $(\tilde{Y}_i^{(k,j)}, \tilde{Z}_i^{(k,j)})$ is a generalized characteristic vector of order k corresponding to $\lambda_i = 0$, then $(-1)^k(\tilde{Y}_i^{(k,j)}, -\tilde{Z}_i^{(k,j)})$ is also a generalized characteristic vector of order k, corresponding to λ_i.

Now we are in a position to give formulae for the general solutions of (5.62) and (5.46). Using the symmetry conditions (5.60) and the properties of generalized characteristic vectors corresponding to characteristic roots λ_i and $-\lambda_i$, we arrive at the following expressions:

$$\tilde{U}(s) = \sum_i \sum_{j=1}^{p_i} \sum_{k=1}^{r_i^{(j)}} \tilde{Y}_i^{(k,j)} \sum_{q=k}^{r_i^{(j)}} \frac{c_{ijq}}{(q-k)!} [s^{q-k} \exp(\lambda_i s) + (h-s)^{q-k} \exp(\lambda_i (h-s))]$$

$$\tag{5.94}$$

$$\tilde{W}(s) = \sum_i \sum_{j=1}^{p_i} \sum_{k=1}^{r_i^{(j)}} Z_i^{(k,j)} \sum_{q=k}^{r_i^{(j)}} \frac{c_{ijq}}{(q-k)!} [s^{q-k} \exp(\lambda_i s) - (h-s)^{q-k} \exp(\lambda_i(h-s))]$$
(5.95)

where the first sum (over index i) contains one term for each pair of non-zero characteristic roots (λ_i, $-\lambda_i$) and one term for the characteristic root equal to zero $\lambda_i = 0$, if such a root exists, c_{ijq} are arbitrary constants. From (5.94) and (5.95) we easily obtain

$$L(s) = \sum_i \sum_{j=1}^{p_i} \sum_{k=1}^{r_i^{(j)}} \sum_{q=k}^{r_i^{(j)}} \frac{c_{ijq}}{(q-k)!} [F_i^{(k,j)} s^{q-k} \exp(\lambda_i s) + H_i^{(k,j)}(h-s)^{q-k} \exp(\lambda_i(h-s))].$$
(5.96)

A practical way of solving the Lyapunov system of equations (5.45)–(5.47) leads through the following steps.

- Calculation of characteristic roots from equation (5.75)
- Determination of characteristic matrices $Y_i^{(k,j)}$ and $Z_i^{(k,j)}$ (or $F_i^{(k,j)}$ and $H_i^{(k,j)}$)
- Construction and solution of the set of linear algebraic equations resulting from (5.45)–(5.47) after the substitution of $L(0)$ to (5.47) and $L(h)$ to (5.45).

5.4 EXAMPLE

In many practical cases a considerable simplification of computation may be achieved by using more or less ingenious modifications of the above algorithm. We shall illustrate this on Example 5.2 from section 5.1. In this case $n = 2$,

$$A = \begin{pmatrix} 0 & 1 \\ 0 & -a \end{pmatrix}, \quad B = \begin{pmatrix} 0 & 0 \\ -b & -c \end{pmatrix}, \quad V = \begin{pmatrix} 0 & 0 \\ 0 & 1 \end{pmatrix}. \quad (5.97)$$

The system of equations (5.62) takes the form

$$\dot{\tilde{U}}_1(s) = b\tilde{W}_2(s)$$
$$\dot{\tilde{U}}_2(s) = \tilde{W}_1(s) - a\tilde{W}_2(s) + b\tilde{W}_4(s)$$
$$\dot{\tilde{U}}_3(s) = c\tilde{W}_2(s)$$
$$\dot{\tilde{U}}_4(s) = \tilde{W}_3(s) - (a-c)\tilde{W}_4(s)$$
$$\dot{\tilde{W}}_1(s) = -b\tilde{U}_2(s) \qquad (5.98)$$
$$\dot{\tilde{W}}_2(s) = \tilde{U}_1(s) - a\tilde{U}_2(s) - b\tilde{U}_4(s)$$
$$\dot{\tilde{W}}_3(s) = -c\tilde{U}_2(s)$$
$$\dot{\tilde{W}}_4(s) = \tilde{U}_3(s) - (a+c)\tilde{U}_4(s)$$

with the initial conditions (5.59)

98   The Lyapunov method   [Ch. 5

$$\tilde{U}_1(0) + \tilde{W}_1(0) = -bM_2$$
$$\tilde{U}_2(0) + \tilde{W}_2(0) = -bM_4$$
$$\tilde{U}_3(0) + \tilde{W}_3(0) = -cM_2$$
$$\tilde{U}_4(0) + \tilde{W}_4(0) = -cM_4$$
(5.99)

and the symmetry conditions (5.60). We have denoted here

$$M = \begin{pmatrix} M_1 & M_3 \\ M_2 & M_4 \end{pmatrix} = M_T.$$

It can be observed that the variables

$$y_1 = c\tilde{U}_1 - b\tilde{U}_3$$
$$y_2 = c\tilde{U}_2 - b\tilde{U}_4$$
$$y_3 = c\tilde{W}_1 - b\tilde{W}_3$$
$$y_4 = c\tilde{W}_2 - b\tilde{W}_4$$

satisfy the following system of differential equations:

$$\dot{y}_1 = \dot{y}_3 = 0$$
$$\dot{y}_2 = y_3 - ay_4$$
$$\dot{y}_4 = y_1 - ay_2$$

with initial conditions

$$y_1(0) + y_3(0) = y_2(0) + y_4(0) = 0.$$

If we take into account the symmetry conditions

$$y_i(s) = y_i(h-s), \quad i = 1,2$$
$$y_i(s) = -y_i(h-s), \quad i = 3,4$$

we easily obtain that $y_i(s) = 0$, $\forall s \in [0,h]$, $i = 1,2,3,4$. Hence

$$c(\tilde{U}_1(s), \tilde{U}_2(s), \tilde{W}_1(s), \tilde{W}_2(s)) = b(\tilde{U}_3(s), \tilde{U}_4(s), \tilde{W}_3(s), \tilde{W}_4(s)), \forall s \in [0, h]. \quad (5.100)$$

Let us now assume $b \neq 0$. After the substitution of (5.100) into (5.98) we obtain

$$\dot{\tilde{U}}_1(s) = b\tilde{W}_2(s)$$
$$\dot{\tilde{U}}_2(s) = \tilde{W}_1(s) - (a-c)\tilde{W}_2(s)$$
$$\dot{\tilde{W}}_1(s) = -b\tilde{U}_2(s)$$
$$\dot{\tilde{W}}_2(s) = \tilde{U}_1(s) - (a+c)\tilde{U}_2(s).$$
(5.101)

Denote by P the matrix of coefficients of the right-hand side, (with the vector of unknowns in a row notation).

$$P = \begin{pmatrix} 0 & 0 & 0 & 1 \\ 0 & 0 & -b & -a-c \\ 0 & 1 & 0 & 0 \\ b & c-a & 0 & 0 \end{pmatrix}. \tag{5.102}$$

The characteristic equation is

$$\det(P - \lambda I) = \lambda^4 + (c^2 - a^2)\lambda^2 - b^2 = 0. \tag{5.103}$$

Hence we obtain four characteristic roots λ_i, $i = 1, \ldots, 4$,

$$\lambda_{1,2} = \pm\sqrt{q}, \quad \lambda_{3,4} = \pm i\sqrt{r} \tag{5.104}$$

where i is the imaginary unit,

$$q = \tfrac{1}{2}[\sqrt{(a^2 - c^2)^2 + 4b^2} + a^2 - c^2]$$
$$r = \tfrac{1}{2}[\sqrt{(a^2 - c^2)^2 + 4b^2} - a^2 + c^2]. \tag{5.105}$$

For each characteristic root λ_i we calculate the corresponding characteristic vector z_i, $i = 1, \ldots, 4$,

$$z_i(P - \lambda_i I) = 0 \tag{5.106}$$

whence, for $a + c \neq 0$,

$$z_i = (b(a+c), b - \lambda_i^2, \frac{b}{\lambda_i}(\lambda_i^2 - b), (a+c)\lambda_i). \tag{5.107}$$

If $a + c = 0$ and $a \neq 0$, we have $q = r = |b|$ and

$$z_i = (b(\lambda_i^2 + b), -2a\lambda_i^2, 2ab\lambda_i, \lambda_i(\lambda_i^2 + b)). \tag{5.108}$$

If $a = c = 0$, we again have $q = r = |b|$ and

$$z_i = (b(\lambda_i^2 + b), b - \lambda_i^2, \lambda_i(b - \lambda_i^2), \lambda_i(\lambda_i^2 + b)). \tag{5.109}$$

After imposing the symmetry requirements on the general solution of (5.101) we obtain

$$\begin{aligned}
\tilde{U}_1(s) &= \alpha_1 z_{11} \cosh\left[\sqrt{q}\left(s - \frac{h}{2}\right)\right] + \alpha_2 z_{31} \cos\left[\sqrt{r}\left(s - \frac{h}{2}\right)\right] \\
\tilde{U}_2(s) &= \alpha_1 z_{12} \cosh\left[\sqrt{q}\left(s - \frac{h}{2}\right)\right] + \alpha_2 z_{32} \cos\left[\sqrt{r}\left(s - \frac{h}{2}\right)\right] \\
\tilde{W}_1(s) &= \alpha_1 z_{13} \sinh\left[\sqrt{q}\left(s - \frac{h}{2}\right)\right] + \alpha_2 \bar{z}_{33} \sin\left[\sqrt{r}\left(s - \frac{h}{2}\right)\right] \\
\tilde{W}_2(s) &= \alpha_1 z_{14} \sinh\left[\sqrt{q}\left(s - \frac{h}{2}\right)\right] + \alpha_2 \bar{z}_{34} \sin\left[\sqrt{r}\left(s - \frac{h}{2}\right)\right]
\end{aligned} \tag{5.110}$$

where $\bar{z}_{33}$ and $\bar{z}_{34}$ are real quantities, $\bar{z}_{33} = iz_{33}$, $\bar{z}_{34} = iz_{34}$ (i is the imaginary unit), α_1 and α_2 are arbitrary real constants, to be determined from boundary conditions.

100 The Lyapunov method [Ch. 5

Let us denote

$$L(s) = \begin{pmatrix} L_1(s) & L_3(s) \\ L_2(s) & L_4(s) \end{pmatrix}. \tag{5.111}$$

From (5.1110) we obtain

$$L_1(s) = \alpha_1 \left\{ z_{11} \cosh\left[\sqrt{q}\left(s - \frac{h}{2}\right)\right] + z_{13} \sinh\left[\sqrt{q}\left(s - \frac{h}{2}\right)\right] \right\}$$

$$+ \alpha_2 \left\{ z_{31} \cos\left[\sqrt{r}\left(s - \frac{h}{2}\right)\right] + \bar{z}_{33} \sin\left[\sqrt{r}\left(s - \frac{h}{2}\right)\right] \right\}$$

$$L_2(s) = \alpha_1 \left\{ z_{12} \cosh\left[\sqrt{q}\left(s - \frac{h}{2}\right)\right] + z_{14} \sinh\left[\sqrt{q}\left(s - \frac{h}{2}\right)\right] \right\}$$

$$+ \alpha_2 \left\{ z_{32} \cos\left[\sqrt{r}\left(s - \frac{h}{2}\right)\right] + \bar{z}_{34} \sin\left[\sqrt{r}\left(s - \frac{h}{2}\right)\right] \right\} \tag{5.112}$$

and from (5.100)

$$L_3(s) = \frac{c}{b} L_1(s) \quad L_4(s) = \frac{c}{b} L_2(s). \tag{5.113}$$

Let

$$Q = \frac{h}{2}\sqrt{q}, \quad R = \frac{h}{2}\sqrt{r}. \tag{5.114}$$

We calculate

$$\left.\begin{array}{l}
L_1(0) = \alpha_1(z_{11} \cosh Q - z_{13} \sinh Q) + \alpha_2(z_{31} \cos R - \bar{z}_{33} \sin R) \\
L_2(0) = \alpha_1(z_{12} \cosh Q - z_{14} \sinh Q) + \alpha_2(z_{32} \cos R - \bar{z}_{34} \sin R) \\
L_1(h) = \alpha_1(z_{11} \cosh Q + z_{13} \sinh Q) + \alpha_2(z_{31} \cos R + \bar{z}_{33} \sin R) \\
L_2(h) = \alpha_1(z_{12} \cosh Q + z_{14} \sinh Q) + \alpha_2(z_{32} \cos R + \bar{z}_{34} \sin R)
\end{array}\right\} \tag{5.115}$$

The initial condition (5.47) yields

$$L_1(0) = -bM_2, \quad L_2(0) = -bM_4 \tag{5.116}$$

and from the matrix equation (5.45) we obtain

$$L_1(h) = 0$$
$$L_2(h) + L_3(h) = aM_2 - M_1 \tag{5.117}$$
$$L_4(h) = aM_4 - M_2 - \tfrac{1}{2}.$$

By using conditions (5.113) we transform equations (5.116) and (5.117) into a set of five linear algebraic equations for five unknowns: α_1, α_2, M_1, M_2 and M_4. This set has the form

$$L_1(h) = 0$$

$$L_1(0) = aL_2(0) - cL_2(h) = \frac{b}{2}$$

$$M_2 = -\frac{1}{b}L_1(0) \qquad (5.118)$$

$$M_4 = -\frac{1}{b}L_2(0)$$

$$M_1 = aM_2 - L_2(h).$$

We find the constants α_1 and α_2 from the first two equations and later we compute M_2, M_4 and M_1 from the three last equations. Define

$$S_i^+ = z_{1i} \cosh Q + z_{1,i+2} \sinh Q, \quad i = 1,2$$
$$S_i^- = z_{1i} \cosh Q - z_{1,i+2} \sinh Q$$
$$T_i^+ = z_{3i} \cos R + \bar{z}_{3,i+2} \sin R$$
$$T_i^- = z_{3i} \cos R - \bar{z}_{3,i+2} \sin R$$
$$\Delta = 2S_1^+[T_1^- - aT_2^- - cT_2^+] - 2T_1^+[S_1^- - aS_2^- - cS_2^+].$$

The solution of system (5.118) can then be written in the form:

$$M_1 = [T_1^+(aS_1^- + bS_2^+) - S_1^+(aT_1^- + bT_2^+)]/\Delta$$
$$M_2 = (T_1^+ S_1^- - S_1^+ T_1^-)/\Delta$$
$$M_4 = (T_1^+ S_2^- - S_1^+ T_2^-)/\Delta \qquad (5.118a)$$
$$\alpha_1 = -bT_1^+/\Delta, \quad \alpha_2 = bS_1^+/\Delta.$$

Function $L(s)$, $s \in [0, h]$, is given by formulae (5.111)–(5.113).

5.5 CASE OF ARBITRARY INITIAL CONDITIONS

In this section we shall consider the evaluation of the performance index (5.5) on trajectories of an asymptotically stable system (5.34) with general initial conditions

$$x(0) = x_0, \quad x(t) = \varphi(t) \quad \text{for } t \in [-h, 0). \qquad (5.119)$$

x_0 is an arbitrary n-dimensional vector and φ is an arbitrary vector-valued and square integrable function, that is, the integral

$$\int_{-h}^{0} \varphi(t)^{\mathrm{T}} \varphi(t) \, \mathrm{d}t \qquad (5.120)$$

exists. The solution of the Cauchy problem (5.34), (5.119) can be expressed by the variation-of-constants formula

102 The Lyapunov method [Ch. 5

$$x(t) = \Phi(t)x_0 + \int_{-h}^{0} \Phi(t - s - h)B\varphi(s) \, ds \tag{5.121}$$

where $\Phi(t)$, the fundamental matrix solution, is determined by equations (5.36). We substitute formula (5.121) into the integral criterion (5.5) and obtain

$$J(x_0, \varphi) = J_a(x_0) + 2J_b(x_0, \varphi) + J_c(\varphi) \tag{5.122}$$

where we use the notation $J(x_0, \varphi)$ to indicate the dependence of J on initial conditions. $J_a(x_0)$, $J_b(x_0, \varphi)$ and $J_c(\varphi)$ are determined by the following expressions:

$$J_a(x_0) = x_0^T \int_0^\infty \Phi(t)^T V \Phi(t) \, dt \, x_0 \tag{5.123}$$

$$J_b(x_0, \varphi) = x_0^T \int_0^\infty \Phi(t)^T V \int_{-h}^{0} \Phi(t - s - h)B\varphi(s) \, ds \, dt \tag{5.124}$$

$$J_c(\varphi) = \int_0^\infty \left[\int_{-h}^{0} \Phi(t - s - h)B\varphi(s) \, ds \right]^T V \left[\int_{-h}^{0} \Phi(t - s - h)B\varphi(s) \, ds \right] dt. \tag{5.125}$$

A comparison of (5.123) with (5.39) shows that

$$J_a(x_0) = x_0^T M x_0. \tag{5.126}$$

Let us consider $J_b(x_0, \varphi)$. By virtue of the asymptotic stability assumption the function under the integral signs is absolutely integrable and we may change the order of integration, which yields

$$J_b(x_0, \varphi) = x_0^T \int_{-h}^{0} \left[\int_0^\infty \Phi(t)^T V \Phi(t - s - h) B \, dt \right] \varphi(s) \, ds$$

$$= x_0^T \int_{-h}^{0} \left[\int_0^\infty \Phi(t + s + h)^T V \Phi(t) B \, dt \right] \varphi(s) \, ds$$

$$= x_0^T \int_{-h}^{0} L(s + h)\varphi(s) \, ds. \tag{5.127}$$

The last expression results from the definition (5.44). The assumption of asymptotic stability allows us to change the order of integration also in the third term of (5.122), which gives

$$J_c(\varphi) = \int_{-h}^{0}\int_{-h}^{0}\int_{0}^{\infty} \varphi(s)^T B^T \Phi(t-s-h)^T V \Phi(t-z-h) B \varphi(z) \, dt \, dz \, ds. \tag{5.128}$$

The integrand $F(s, z)$ in (5.128) is a symmetric function, that is, $F(s, z) = F(z, s)$. Hence

$$J_c(\varphi) = 2 \int_{-h}^{0}\int_{s}^{0} \varphi(s)^T B^T \left[\int_{0}^{\infty} \Phi(t-s-h)^T V \Phi(t-z-h) B \, dt\right] \varphi(z) \, dz \, ds$$

$$= 2 \int_{-h}^{0}\int_{s}^{0} \varphi(s)^T B^T \left[\int_{0}^{\infty} \Phi(t-s+z)^T V \Phi(t) B \, dt\right] \varphi(z) \, dz \, ds$$

$$= 2 \int_{-h}^{0}\int_{s}^{0} \varphi(s)^T B^T L(z-s) \varphi(z) \, dz \, ds. \tag{5.129}$$

Equivalently,

$$J_c(\varphi) = \int_{-h}^{0}\int_{-h}^{0} \varphi(s)^T L_1(s, z) \varphi(z) \, dz \, ds \tag{5.130}$$

where

$$L_1(s, z) = \begin{cases} L(s-z)^T B & \text{for } z \leq s \\ B^T L(z-s) & \text{for } z > s \end{cases}. \tag{5.131}$$

As a simple example, let us consider the case where the initial condition is a constant function in all of the interval $[-h, 0]$. Thus $x_0 = \varphi(s) = a$, $\forall s \in [-h, 0]$ for some a. We substitute this into (5.126), (5.127) and (5.129) whence

$$J_a(x_0) = a^T M a \tag{5.132}$$

$$J_b(x_0, \varphi) = a^T \int_{0}^{h} L(s) \, ds \, a \tag{5.133}$$

$$J_c(\varphi) = 2 a^T B^T \int_{0}^{h} (h-s) L(s) \, ds \, a. \tag{5.134}$$

Let us apply these formulae to Example 5.1. We obtain

$$J_a(x_0) = \frac{1 + \sin Kh}{2K \cos Kh} a^2 \tag{5.135}$$

$$J_b(x_0, \varphi) = \int_0^h \frac{\sin K(s-h) - \cos Ks}{2 \cos Kh} \, ds \, a^2$$

$$= \frac{\cos Kh - \sin Kh - 1}{2K \cos Kh} a^2 \qquad (5.136)$$

$$J_c(\varphi) = \frac{a^2 K}{\cos Kh} \int_0^h (s-h)[\sin K(s-h) - \cos Ks] \, ds$$

$$= \frac{a^2}{K \cos Kh} [\sin Kh - (Kh+1)\cos Kh + 1] \qquad (5.137)$$

and finally, according to (5.122),

$$J(x_0, \varphi) = \frac{a^2}{2K \cos Kh} (\sin Kh - 2Kh \cos Kh + 1). \qquad (5.138)$$

5.6 PERFORMANCE CRITERIA WITH WEIGHTING FUNCTIONS AND DERIVATIVES

In some applications the performance of a system is evaluated by means of integral criteria, more general than that considered in the previous sections. Of special interest are criteria in which the integrand in (5.5) is multiplied by a function of time, called a weighting function. It often happens that the behaviour of a control system is not equally important at every moment of time. A natural way of representing these differences in importance is to multiply the integrand in the performance criterion (5.5) by a non-negative function: the more important is the time moment, the greater is the value of the weighting function. This is a consequence of the assumption that smaller values of the criterion correspond to better performance of the system. When optimizing a control system by means of minimization of a performance index with a weighting function, we obtain a relatively better system performance on the intervals of time where the weighting function attains large values. Of course, this is at the cost of a relative deterioration on the intervals with smaller values of the weighting function.

In many control systems the rate of convergence of some variables to their steady-state values (or to the equilibrium point) is a crucial factor. In order to make the integral (5.5) a good tool for the estimation of performance of such systems, we multiply the integrand by a rapidly increasing function, e.g. t^k for some positive integer k. We obtain a more precise and universal tool, if the weighting function is of the form $t^k \exp(at)$. The exponential factor allows us to estimate most precisely the rate with which some variables tend to zero, as well as to assign a greater weight to a selected interval of time.

These remarks justify our interest in the performance criterion of the form

$$J_{a,k} = \int_0^\infty t^k \exp(at) x(t)^T V x(t) \, dt \tag{5.139}$$

or, more generally,

$$J = \int_0^\infty x(t)^T V(t) x(t) \, dt \tag{5.140}$$

where

$$V(t) = \sum_{i=0}^k V_i t^i \exp(a_i t). \tag{5.141}$$

V and V_i, $i = 0, \ldots, k$, are symmetric and positive semidefinite matrices, $x(t)$ is a trajectory of system (5.34). We begin our considerations with the performance index (5.139) for the case $k = 0$. We may then write

$$J_{a,0} = \int_0^\infty y(t)^T V y(t) \, dt \tag{5.142}$$

where

$$y(t) = \exp\left(\frac{at}{2}\right) x(t). \tag{5.143}$$

We therefore have a differential equation for y,

$$\dot{y}(t) = \left(A + \frac{a}{2} I\right) y(t) + \exp\left(\frac{ah}{2}\right) B y(t - h). \tag{5.144}$$

Let us note that the characteristic equation for (5.144) is

$$\det\left[\left(\lambda - \frac{a}{2}\right) I - A - B \exp\left(-\lambda + \frac{a}{2}\right) h\right] = 0. \tag{5.145}$$

Hence it is evident that the characteristic roots of system (5.144) differ from the characteristic roots of system (5.34) by a constant shift equal to $a/2$, that is, λ is a characteristic root of (5.144) if and only if $\lambda - a/2$ is a characteristic root of (5.34).

We assume that integral (5.142) exists. A sufficient condition for this is that system (5.144) is asymptotically stable, i.e. all characteristic roots of (5.145) have negative real parts. The Lyapunov system of equations (5.45)–(5.47) for system (5.144) has the form

$$M_a A + A^T M_a + a M_a + V + L_a(h)^T + L_a(h) = 0 \tag{5.146}$$

$$\dot{L}_a(s) = A^T L_a(s) + \frac{a}{2} L_a(s) + L_a(h-s)^T B \exp\left(\frac{ah}{2}\right) \tag{5.147}$$

$$L_a(0) = M_a B \exp\left(\frac{ah}{2}\right). \tag{5.148}$$

The subscript in M_a and L_a denotes the dependence on the parameter a. This set of equations can be studied and solved along the lines of section 5.3, after replacing A and B in (5.45)–(5.47) by $A + aI/2$ and $B \exp(ah/2)$, respectively.

We shall now pass to the case $k > 0$. Let us observe that integral (5.139) is a differentiable function of the parameter a. Moreover, derivatives of all orders exist and are continuous functions of a. Since the integrand is absolutely integrable, the differentiation may be performed under the integral sign, which yields

$$\frac{d^k}{da^k} J_{a,0} = \int_0^\infty \frac{d^k}{da^k} \exp(at) x(t)^T V x(t) \, dt$$

$$= \int_0^\infty t^k \exp(at) x(t)^T V x(t) \, dt. \tag{5.149}$$

Thus

$$J_{a,k} = \frac{d^k}{da^k} J_{a,0}. \tag{5.150}$$

In particular, if we want to calculate the performance index

$$J_{0,k} = \int_0^\infty t^k x(t)^T V x(t) \, dt \tag{5.151}$$

we take the limit of the right-hand side of (5.150) at $a = 0$, that is,

$$J_{0,k} = \lim_{a \to 0} \frac{d^k}{da^k} J_{a,0} = \frac{d^k}{da^k} J_{a,0}\big|_{a=0}. \tag{5.152}$$

The relationships (5.150) and (5.152) offer a convenient way of computing the performance indices $J_{a,k}$ for any admissible a, provided we have at our disposal a finite analytic formula for $J_{a,0}$ which shows an explicit, easily differentiable dependence on a. However, this is so only in simplest problems with time delays. In more complicated cases it might be preferable to use a recursive system of Lyapunov-type equations which we shall now derive. Assume that x satisfies the initial conditions $x(0) = x_0$, $x(t) = \varphi(t)$ for $t < 0$. By virtue of formulae (5.122), (5.126), (5.127) and (5.129) we have

$$J_{a,0} = x_0^T M_a x_0 + 2 x_0^T \int_{-h}^0 L_a(s+h) \exp\left(\frac{as}{2}\right) \varphi(s) \, ds$$

$$+ 2 \int_{-h}^{0} \int_{s}^{0} \varphi(s)^T B^T L_a(z-s) \exp\left(\frac{a}{2}(z+s+h)\right) \varphi(z) \, dz \, ds. \tag{5.153}$$

Denote by $M_a^{(k)}$ and $L_a^{(k)}(s)$ the kth derivatives with respect to a of M_a and $L_a(s)$, respectively. Since all the integrands in (5.153) are absolutely integrable, we obtain from (5.150) and (5.153)

$$J_{a,k} = x_0^T M_a^{(k)} x_0 + 2 x_0^T \int_{-h}^{0} \frac{\partial^k}{\partial a^k} \left[L_a(s+h) \exp\left(\frac{as}{2}\right) \right] \varphi(s) \, ds$$

$$+ 2 \int_{-h}^{0} \int_{s}^{0} \varphi(s)^T B^T \frac{\partial^k}{\partial a^k} \left[L_a(z-s) \exp\left(\frac{a}{2}(z+s+h)\right) \right] \varphi(z) \, dz \, ds.$$

$$\tag{5.154}$$

By differentiating the equalities (5.146)–(5.148) k times with respect to a, we get for $k > 0$

$$M_a^{(k)} A + A^T M_a^{(k)} + a M_a^{(k)} + k M_a^{(k-1)} + L_a^{(k)}(h)^T + L_a^{(k)}(h) = 0 \tag{5.155}$$

$$\dot{L}_a^{(k)}(s) = A^T L_a^{(k)}(s) + \frac{a}{2} L_a^{(k)}(s) + \frac{k}{2} L_a^{(k-1)}(s)$$

$$+ \frac{\partial^k}{\partial a^k} \left[\exp\left(\frac{ah}{2}\right) L_a(h-s)^T \right] B \tag{5.156}$$

$$L_a^{(k)}(0) = \frac{\partial^k}{\partial a^k} \left[\exp\left(\frac{ah}{2}\right) M_a \right] B. \tag{5.157}$$

These equations, together with (5.146)–(5.148), form a system which can be solved in a recursive way. From (5.146)–(5.148) we calculate $M_a = M_a^{(0)}$ and $L_a = L_a^{(0)}$. This allows us to solve (5.155)–(5.157) for $k = 1$ which gives $M_a^{(1)}$ and $L_a^{(1)}$. We can then solve (5.155)–(5.157) for $k = 2$, and so on for increasing values of k. Notice that

$$\frac{\partial^k}{\partial a^k} \left[\exp\left(\frac{ah}{2}\right) L_a(s) \right] = \exp\left(\frac{ah}{2}\right) \sum_{i=0}^{k} \binom{k}{i} \left(\frac{h}{2}\right)^{k-i} L_a^{(i)}(s) \tag{5.158}$$

and similarly

$$\frac{\partial^k}{\partial a^k} \left[\exp\left(\frac{ah}{2}\right) M_a \right] = \exp\left(\frac{ah}{2}\right) \sum_{i=0}^{k} \binom{k}{i} \left(\frac{h}{2}\right)^{k-i} M_a^{(i)}. \tag{5.159}$$

The only essential difference between system (5.155)–(5.157) and system (5.146)–(5.148) results from the fact that equation (5.156) is non-homogeneous. Indeed, if we denote

$$f_k(s) = \frac{k}{2} L_a^{(k-1)}(s) + \exp\left(\frac{ah}{2}\right) \sum_{i=0}^{k-1} \binom{k}{i} \left(\frac{h}{2}\right)^{k-1} L_a^{(i)}(h-s)^T B \tag{5.160}$$

we can rewrite (5.156) in the form

$$\dot{L}_a^{(k)}(s) = \left(A + \frac{a}{2}I\right)^T L_a^{(k)}(s) + L_a^{(k)}(h-s)^T B \exp\left(\frac{ah}{2}\right) + f_k(s). \qquad (5.161)$$

We shall briefly show how the results of section 5.3 can be applied to solve equation (5.161). $L_a^{(k)}$ is transformed into a pair of matrix functions U, W, according to (5.56). These functions satisfy the following set of differential equations:

$$\begin{aligned} \dot{U}(s) &= A_a^T W(s) - W(s)^T B_a + w(s) \\ \dot{W}(s) &= A_a^T U(s) + U(s)^T B_a + u(s) \end{aligned} \qquad (5.162)$$

where

$$A_a = A + \frac{a}{2}I, \quad B_a = B \exp\left(\frac{ah}{2}\right)$$

$$w(s) = \tfrac{1}{2}[f_k(s) - f_k(h-s)]$$

$$u(s) = \tfrac{1}{2}[f_k(s) + f_k(h-s)], \qquad (5.163)$$

with the boundary conditions

$$U(0) + W(0) = U(h) - W(h) = \frac{\partial^k}{\partial a^k}\left[\exp\left(\frac{ah}{2}\right)M_a\right]B. \qquad (5.164)$$

The symmetry conditions (5.60) are also satisfied. After introducing the vectors $\tilde{U}$ and $\tilde{W}$ and decomposing u and w in the same manner,

$$\begin{aligned} \tilde{u} &= (u^{(1)^T}, u^{(2)^T}, \ldots, u^{(n)^T}) \\ \tilde{w} &= (w^{(1)^T}, w^{(2)^T}, \ldots, w^{(n)^T}) \end{aligned} \qquad (5.165)$$

we obtain the following vector equation

$$(\dot{\tilde{U}}(s), \dot{\tilde{W}}(s)) = (\tilde{U}(s), \tilde{W}(s))N_a + (\tilde{w}(s), \tilde{u}(s)). \qquad (5.166)$$

N_a is defined by (5.63) with A and B replaced by A_a and B_a, respectively. The solution of this equation has the form

$$(\tilde{U}(s), \tilde{W}(s)) = \alpha \exp(N_a s) + \int_0^s (\tilde{w}(z), \tilde{u}(z)) \exp[N_a(s-z)] \, dz \qquad (5.167)$$

where α is a row vector of dimension $2n^2$, to be determined from the boundary conditions (5.164) and the symmetry conditions (5.60), which can be done similarly as in section 5.3.

To finish the discussion of performance criteria with weighting functions, let us observe that the system of equations (5.155)–(5.157) can also be used to determine the value of $J_{0,k}$; one only has to compute all the derivatives with respect to a at $a = 0$.

We shall now discuss the integral performance criterion of the form

$$J = \int_0^\infty \dot{x}(t)^T V \dot{x}(t)\, dt \tag{5.168}$$

where $\dot{x}(t)$ denotes, as usually, the time derivative of $x(t)$, and $x(t)$ is a trajectory of system (5.34). V is a symmetric and positive semidefinite matrix. We can justify our interest in this criterion by recalling that in many cases a good control system should have 'smooth' transients, that is, without many rapid changes. A criterion with squared derivatives takes small values for smooth processes, and large, for processes changing rapidly with high frequency. This is important in parametric optimization of control systems. If we minimize (by an appropriate choice of parameters) a performance index which is a linear combination of terms of the form (5.5) and (5.168), we are able to obtain a system with smooth and rapidly vanishing transients. It should be stressed, however, that often it is possible to include implicitly terms with squared derivatives, of the first or higher order, into the performance index (5.5). This is due to the fact that some components of the vector $x(t)$ may be defined as equal to derivatives of other components. This also explains why it is hardly necessary to consider terms with quadratic functions of higher derivatives of x.

It is easy to see that the performance index (5.168) can be written in the form

$$J = \dot{x}(0)^T M \dot{x}(0) + 2\dot{x}(0)^T \int_{-h}^{0} L(s+h)\dot{x}(s)\, ds$$

$$+ 2 \int_{-h}^{0} \int_{s}^{0} \dot{x}(s)^T B^T L(z-s)\dot{x}(z)\, dz\, ds, \tag{5.169}$$

if the trajectory x is continuously differentiable for every $t > -h$ and J exists. Asymptotic stability of system (5.34) is sufficient for the existence of J. The trajectory is continuously differentiable if the initial function $x:[-h, 0] \to \mathbb{R}^n$ is continuously differentiable and its left-hand-side derivative at zero is equal to $Ax(0) + Bx(-h)$. The matrix M and the matrix function L are the solution of the Lyapunov system of equations (5.45)–(5.47), that is, the Lyapunov system corresponding to criterion (5.5).

The substitution $\dot{x}(0) = Ax(0) + Bx(-h)$ and integration by parts of the terms involving derivatives $\dot{x}(s)$ allows us to rewrite (5.169) as follows:

$$J = x(0)^T [A^T M A + A^T L(h) + \dot{L}(h)] x(0)$$

$$- 2x(0)^T \int_{-h}^{0} \ddot{L}(s+h) x(s)\, ds - 2 \int_{-h}^{0} x(s)^T B^T \dot{L}(0) x(s)\, ds$$

$$- 2 \int_{-h}^{0} \int_{s}^{0} x(s)^T B^T \ddot{L}(z-s) x(z)\, dz\, ds. \tag{5.170}$$

The first- and second-order derivatives of L, denoted by $\dot{L}$ and $\ddot{L}$ respectively, can be readily computed from equation (5.46).

Although expression (5.170) was derived for a special class of initial conditions, it is valid for every initial condition from the space $H = \mathbb{R}^n \times L^2(-h, 0; \mathbb{R}^n)$. Indeed, J in (5.168) is a continuous function of initial conditions from H. The set of all initial conditions which generate a continuously differentiable trajectory $x(t)$ for all $t > -h$ is dense in H. Now, (5.170) has a unique continuous extension onto the whole space H, which ends the proof.

5.7 SYSTEMS WITH MANY DELAYS

The Lyapunov method may be also applied to systems with many time delays in the equation describing system dynamics

$$\dot{x}(t) = \sum_{i=0}^{k} A_i x(t - h_i) \tag{5.171}$$

where $x(t)$, as usually, is an n-dimensional column vector, A_i, $i = 0, 1, \ldots, k$, are $n \times n$ constant matrices and h_i are time delays, $0 = h_0 < h_1 < \ldots < h_k$. However, as elsewhere, finite analytic formulae for the integral performance index (5.5) can be derived only in the case of commensurate delays, that is, if

$$h_i = ih, \quad i = 0, 1, \ldots, k, \tag{5.172}$$

for some positive h. We therefore confine ourselves to commensurate delays.

If we assume the initial conditions in the form

$$x(0) = x_0, \quad x(t) = \varphi(t) \quad \text{for } t < 0 \tag{5.173}$$

where x_0 is an arbitrary vector in $\mathbb{R}^n$ and $\varphi : [-h_k, 0) \to \mathbb{R}^n$ is an arbitrary square integrable function, the solution of (5.171) can be written in the form

$$x(t) = \Phi(t) x_0 + \sum_{i=1}^{k} \int_{-h_i}^{0} \Phi(t - s - h_i) A_i \varphi(s) \, ds \tag{5.174}$$

where $\Phi(t)$ is the fundamental matrix solution, satisfying the differential equation

$$\dot{\Phi}(t) = \sum_{i=0}^{k} A_i \Phi(t - h_i), \quad t \geq 0$$

$$\Phi(0) = I, \quad \Phi(t) = 0 \text{ for } t < 0. \tag{5.175}$$

The following identity, analogous to (5.41), can be readily verified by Laplace-transform techniques

$$\sum_{i=0}^{k} A_i \Phi(t - h_i) = \sum_{i=0}^{k} \Phi(t - h_i) A_i, \quad \forall t. \tag{5.176}$$

In the derivation of the Lyapunov system of equations we shall follow the lines of reasoning of section 5.2. Similarly as we did there, we shall simplify our considerations by assuming $\varphi = 0$. We then have for the performance criterion (5.5)

$$J = x_0^T M x_0 \tag{5.177}$$

where

$$M = \int_0^\infty \Phi(t)^T V \Phi(t) \, dt$$

$$= \int_s^\infty \Phi(t-s)^T V \Phi(t-s) \, dt, \ \forall s. \tag{5.178}$$

For the purposes of this section, we define the function L in a slightly different way:

$$L(s) = \int_0^\infty \Phi(t+s)^T V \Phi(t) \, dt, \quad s \in [0, h_k]. \tag{5.179}$$

The differentiation of identities (5.178) and (5.179) with respect to s and some algebraic manipulation yield the following Lyapunov system of equations:

$$V + \sum_{i=0}^k [A_i^T L(h_i)^T + L(h_i) A_i] = 0 \tag{5.180}$$

$$\dot{L}(s) = \sum_{i=0}^k A_i^T \begin{cases} L(h_i - s)^T & \text{for } s - h_i < 0 \\ L(s - h_i) & \text{for } s - h_i \geq 0, \end{cases} \quad s \in [0, h_k] \tag{5.181}$$

$$L(0) = M. \tag{5.182}$$

These equations can be effectively solved with the use of the method described in section 5.3, if the delays are commensurate, that is, (5.172) holds. We shall apply the well-known step method to equation (5.181). Define

$$L_i(s) = L(s + (i-1)h), \quad s \in [0, h], \quad i = 1, \ldots, k. \tag{5.183}$$

This allows us to transform equations (5.180)–(5.182) into the following, equivalent, set of equations:

$$A_0^T M + M A_0 + V + \sum_{i=1}^k [A_i^T L_i(h)^T + L_i(h) A_i] = 0 \tag{5.184}$$

$$\dot{L}_i(s) = \sum_{j=0}^{i-1} A_j^T L_{i-j}(s) + \sum_{j=i}^k A_j^T L_{j-i+1}(h-s)^T, \quad s \in [0, h], \quad i = 1, \ldots, k \tag{5.185}$$

$$L_1(0) = M \tag{5.186}$$

$$L_i(h) = L_{i+1}(0), \quad i = 1, \ldots, k-1. \tag{5.187}$$

To each variable L_i we assign a pair of matrix functions U_i, W_i, similarly as we did in (5.56)

$$U_i(s) = \tfrac{1}{2}[L_i(s) + L_i(h-s)]$$
$$W_i(s) = \tfrac{1}{2}[L_i(s) - L_i(h-s)]. \qquad (5.188)$$

By substituting this into equalities (5.185)–(5.187) we obtain

$$\dot{U}_i(s) = \sum_{j=0}^{i-1} A_j^T W_{i-j}(s) - \sum_{j=i}^{k} A_j^T W_{j-i+1}(s)^T, \quad s\in[0,h], \quad i=1,\ldots,k \qquad (5.189)$$

$$\dot{W}_i(s) = \sum_{j=0}^{i-1} A_j^T U_{i-j}(s) + \sum_{j=i}^{k} A_j^T U_{j-i+1}(s)^T, \quad s\in[0,h], \quad i=1,\ldots,k \qquad (5.190)$$

$$U_1(0) + W_1(0) = M \qquad (5.191)$$

$$U_{i+1}(0) + W_{i+1}(0) = U_i(h) + W_i(h), \quad i=1,\ldots,k-1. \qquad (5.192)$$

Of course, the variables U_i and W_i satisfy the symmetry conditions

$$U_i(s) = U_i(h-s), \quad W_i(s) = -W_i(h-s), \quad i=1,\ldots,k. \qquad (5.193)$$

Now, if we construct vectors of dimension kn^2

$$\tilde{U} = (U_1^{(1)^T}, U_1^{(2)^T}, \ldots, U_1^{(n)^T}, U_2^{(1)^T}, \ldots, U_2^{(n)^T}, \ldots, U_k^{(n)^T})$$
$$\tilde{W} = (W_1^{(1)^T}, W_1^{(2)^T}, \ldots, W_1^{(n)^T}, W_2^{(1)^T}, \ldots, W_2^{(n)^T}, \ldots, W_k^{(n)^T}) \qquad (5.194)$$

we can easily rewrite system (5.189)–(5.192) as a system of $2kn^2$ ordinary linear differential equations with two-sided boundary conditions, whose solution satisfies the symmetry conditions

$$\tilde{U}(s) = \tilde{U}(h-s), \quad \tilde{W}(s) = -\tilde{W}(h-s). \qquad (5.195)$$

We thus obtain a problem which can be solved along the lines of section 5.3. Let us notice that owing to the step method, the interval of integration of the differential equation decreased k times (h in (5.189), (5.190) instead of kh in (5.181)) and the dimension of the system of differential equations increased k times (n^2 in (5.181) and kn^2 in (5.185)).

5.8 NEUTRAL SYSTEMS

Consider a system whose dynamics is described by the neutral equation

$$\sum_{i=0}^{N} [A_i \dot{x}(t-h_i) + B_i x(t-h_i)] = f(t), \quad t>0 \qquad (5.196)$$

where $x(t)$ is an n-dimensional column vector, $\dot{x}(t)$ denotes the derivative of x (at least for positive arguments), the time delays h_i satisfy

$$0 = h_0 < h_1 < h_2 < \ldots < h_N, \tag{5.197}$$

A_i and B_i are constant $n \times n$ matrices and

$$\det A_0 \neq 0, \tag{5.198}$$

f is an n-dimensional, integrable vector function which may represent external inputs or disturbances. The initial conditions for equation (5.196) are

$$x(0) = x_0, \quad x(t) = \varphi(t) \quad \text{for } t \in [-h_N, 0) \tag{5.199}$$

$$\dot{x}(t) = \mu(t), \quad t \in [-h_N, 0). \tag{5.200}$$

In most applications, φ is differentiable in $[-h_N, 0)$, $x_0 = \varphi(0-)$ and μ is equal to the first derivative of φ, but this is not obligatory. It is sufficient to assume that μ and φ are arbitrary integrable functions.

By a solution of the initial value problem (5.196), (5.199), (5.200) we mean a function $x: [-h_N, \infty) \to \mathbb{R}^n$, continuous in $(0, \infty)$ and right-continuous at zero, which satisfies conditions (5.199) and for almost every $t > 0$ fulfils equation (5.196) with the substitution of (5.199) and (5.200) for $x(t)$ and $\dot{x}(t)$ with negative values of t.

We shall need a representation of the solution of (5.196) by means of definite integrals, that is, a variation-of-constants formula analogous to (5.121) or (5.174). It can be shown that under our assumptions the initial value problem (5.196), (5.199), (5.200) has a unique solution and this solution has the following representation. Let us first define the fundamental matrix solution Φ of equation (5.196) as the unique $n \times n$ matrix function which satisfies the following requirements:

(i) $\Phi(0) = A_0^{-1}$, $\Phi(t) = 0$ for $t < 0$ \hfill (5.201)

(ii) the function

$$t \mapsto \sum_{i=0}^{N} A_i \Phi(t - h_i) \tag{5.202}$$

is continuous in $(0, \infty)$ and right-continuous at zero,

(iii) Φ is differentiable almost everywhere in $(0, \infty)$ and for almost every t satisfies the equation

$$\sum_{i=0}^{N} [A_i \dot{\Phi}(t - h_i) + B_i \Phi(t - h_i)] = 0. \tag{5.203}$$

An important property of the fundamental solution Φ is that the function

$$t \mapsto \sum_{i=0}^{N} \Phi(t - h_i) A_i \tag{5.204}$$

has the same continuity properties as (5.202) and

$$\sum_{i=0}^{N} [\dot{\Phi}(t - h_i) A_i + \Phi(t - h_i) B_i] = 0 \tag{5.205}$$

for almost all ts in $(0, \infty)$.

The variation-of-constants formula for equation (5.196) has the form

$$x(t) = \sum_{i=0}^{N} \Phi(t - h_i)A_i x_0 + \int_0^t \Phi(t - s)f(s)\,ds$$

$$- \sum_{i=0}^{N} \int_{-h_i}^{0} \Phi(t - h_i - s)[A_i \mu(s) + B_i \varphi(s)]\,ds. \qquad (5.206)$$

We shall calculate the integral performance criterion (5.5) on trajectories of system (5.196). Let us first assume that the system is homogeneous, that is, $f = 0$ and the initial conditions are pointwise:

$$\varphi = 0 \quad \text{and} \quad \mu = 0.$$

We then have

$$x(t) = \sum_{i=0}^{N} \Phi(t - h_i)A_i x_0. \qquad (5.207)$$

Hence, the integral performance criterion (5.5) takes the form

$$J(x_0) = x_0^T M x_0 \qquad (5.208)$$

where

$$M = \int_0^\infty \psi(t)^T V \psi(t)\,dt \qquad (5.209)$$

$$\psi(t) = \sum_{i=0}^{N} \Phi(t - h_i)A_i. \qquad (5.210)$$

For every real s we have

$$M = \int_s^\infty \psi(t - s)^T V \psi(t - s)\,dt. \qquad (5.211)$$

We differentiate both sides of this equality with respect to s and substitute $s = 0$ which gives

$$0 = V + \int_0^\infty \dot{\psi}(t)^T V \psi(t)\,dt + \int_0^\infty \psi(t)^T V \dot{\psi}(t)\,dt. \qquad (5.212)$$

By virtue of (5.205)

$$\dot{\psi}(t) = - \sum_{i=0}^{N} \Phi(t - h_i)B_i \qquad (5.213)$$

and so we obtain from (5.212)

$$\sum_{i,j=0}^{N} [B_i^T \int_0^\infty \Phi(t-h_i)^T V \Phi(t-h_j) \, dt \, A_j$$

$$+ A_j^T \int_0^\infty \Phi(t-h_j)^T V \Phi(t-h_i) \, dt \, B_i] = V. \tag{5.214}$$

We define

$$L(s) = \int_0^\infty \Phi(t+s)^T V \Phi(t) \, dt \tag{5.215}$$

for every real s. Hence

$$\int_0^\infty \Phi(t-h_i)^T V \Phi(t-h_j) \, dt = \begin{cases} L(h_j - h_i) & \text{for } h_j \geq h_i \\ L(h_i - h_j)^T & \text{for } h_i \geq h_j \end{cases} \tag{5.216}$$

and equation (5.214) yields

$$\sum_{i=0}^{N} [B_i^T L(0) A_i + A_i^T L(0) B_i]$$

$$+ \sum_{i=0}^{N-1} \sum_{j=i+1}^{N} [B_i^T L(h_j - h_i) A_j + A_j^T L(h_j - h_i)^T B_i$$

$$+ B_j^T L(h_j - h_i)^T A_i + A_i^T L(h_j - h_i) B_j] = V. \tag{5.217}$$

We shall now derive a functional–differential equation for the function L. From the definitions (5.215) and (5.210) we easily obtain

$$\sum_{i=0}^{N} A_i^T L(s - h_i) = \int_0^\infty \psi(t+s)^T V \Phi(t) \, dt. \tag{5.218}$$

We differentiate both sides of this equality with respect to s and use (5.205), which gives

$$\frac{d}{ds} \sum_{i=0}^{N} A_i^T \begin{cases} L(s - h_i) & \text{for } s \geq h_i \\ L(h_i - s)^T & \text{for } s < h_i \end{cases}$$

$$+ \sum_{i=0}^{N} B_i^T \begin{cases} L(s - h_i) & \text{for } s \geq h_i \\ L(h_i - s)^T & \text{for } s < h_i \end{cases} = 0, \quad s \in [0, h_N]. \tag{5.219}$$

The differential equation (5.219) together with the boundary condition (5.217) and the obvious symmetry condition

$$L(0) = L(0)^T \geq 0 \tag{5.220}$$

can be used to calculate the function $L(s)$, $s \in [0, h_N]$, in a similar manner as equations (5.45)–(5.47) or equations (5.180)–(5.182). By analogy with the previous cases we call system (5.217), (5.219), (5.220) the Lyapunov set of equations corresponding to problem (5.196), (5.5). We then have by virtue of (5.209) and (5.210)

$$M = \sum_{i=0}^{N} A_i^T L(0) A_i + \sum_{i=0}^{N-1} \sum_{j=i+1}^{N} [A_j^T L(h_j - h_i)^T A_i$$
$$+ A_i^T L(h_j - h_i) A_j]. \quad (5.221)$$

Let us discuss the boundary problem (5.217), (5.219), (5.220) in more detail. First, we note that the function $L(s)$, $s \in [0, h_N]$, is continuous and almost everywhere differentiable. The same properties are possessed by every function of the form

$$[0, h_N] \ni s \mapsto \begin{cases} L(s - h_i) & \text{for } s \geq h_i \\ L(h_i - s)^T & \text{for } s < h_i \end{cases}, \quad i = 0, 1, \ldots, N. \quad (5.222)$$

Therefore, the first sum in the left-hand side of (5.219) is a continuous function, term-by-term differentiable for almost every s. It is easy to see that the existence of integral (5.215) at $s = 0$ is a sufficient condition for the existence of a solution of the Lyapunov set of equations (5.217), (5.219), (5.220). And in turn, the integral

$$\int_0^\infty \Phi(t)^T V \Phi(t) \, dt$$

exists if all roots λ_i of the characteristic equation

$$\det \left[\sum_{i=0}^{N} (\lambda A_i + B_i) \exp(-\lambda h_i) \right] = 0 \quad (5.223)$$

have real parts less than some negative real

$$\operatorname{Re} \lambda_i < \alpha, \, \forall i \quad \text{where } \alpha < 0. \quad (5.224)$$

Condition (5.224) also implies that problem (5.217), (5.219), (5.220) has a unique solution.

It is possible, at least in principle, to express the solution of the Lyapunov set of equations by means of finite, explicit formulae constructed of elementary functions, provided the delays $h_1, h_2, \ldots, h_N$ are commensurate. In this case we may use the method of steps and proceed along the lines of section 5.7.

We shall now pass on to the general case where the solution is given by the full formula (5.206) and we shall find the corresponding expression for the integral performance criterion. Let us first notice that the Lyapunov approach is applicable to problems in which the function f (see (5.196)) is related in a special way to the performance criterion. To see this, consider a criterion of the form

$$J = \int_0^\infty [x(t) - f_1(t)^T] V [x(t) - f_1(t)] \, dt. \quad (5.225)$$

The functions f and f_1 must be related in such a way that the change of variables

$$x(t) = y(t) + f_1(t) \tag{5.226}$$

transforms the non-homogeneous system equation into a homogeneous one. In conclusion, it suffices to assume $f = 0$ and consider the performance index (5.5). We shall write $J(x_0, \mu, \varphi)$ instead of J to stress the dependence on all initial conditions. We have

$$J(x_0, \mu, \varphi) = J_a(x_0) - 2J_b(x_0, \mu, \varphi) + J_c(\mu, \varphi) \tag{5.227}$$

where

$$J_a(x_0) = x_0^T M x_0 \tag{5.228}$$

$$J_b(x_0, \mu, \varphi) = x_0^T \int_0^\infty \psi(t)^T V y(t) \, dt \tag{5.229}$$

$$J_c(\mu, \varphi) = \int_0^\infty y(t)^T V y(t) \, dt \tag{5.230}$$

$$y(t) = \sum_{i=1}^N \int_{-h_i}^0 \Phi(t - h_i - s)[A_i \mu(s) + B_i \varphi(s)] \, ds. \tag{5.231}$$

Similarly as in section 5.5, J_b and J_c can be expressed in terms of the solution of the Lyapunov set of equations. Based on the assumption of asymptotic stability we easily arrive at the following formulae:

$$J_b(x_0, \mu, \varphi) = x_0^T \sum_{i=0}^N \sum_{j=1}^N A_i^T \int_{-h_j}^0 L_{ij}(s) y_j(s) \, ds \tag{5.232}$$

$$J_c(\mu, \varphi) = \sum_{i,j=1}^N \int_{-h_i}^0 \int_{-h_j}^0 y_i(s)^T L_{ij}(\sigma - s) y_j(\sigma) \, d\sigma \, ds \tag{5.233}$$

where

$$L_{ij}(s) = \begin{cases} L(s - h_i + h_j) & \text{for } s \geqslant h_i - h_j \\ L(h_i - h_j - s)^T & \text{for } s < h_i - h_j \end{cases} \tag{5.234}$$

$$y_i(s) = A_i \mu(s) + B_i \varphi(s). \tag{5.235}$$

As we have already mentioned, in most applications it is natural to assume that $\varphi(s)$, $s \in [-h_n, 0)$ is a continuous and differentiable function, $\varphi(0-) = x_0$ and $\mu(s) = d\varphi(s)/ds$, $s \in (-h_N, 0)$. If these assumptions are taken, an appropriate integration by parts may be performed in the variation-of-constants formula (5.206) and in formulae (5.232) and (5.233) yielding new expressions which do not include the derivative $\mu = \dot{\varphi}$. We shall study this approach in the case of one delay, $N = 1$ and $h = h_N$. In this case, the Lyapunov set of equations takes a particularly simple form. From equation (5.217) we obtain

$$A_0^T L(0) B_0 + B_0^T L(0) A_0 + A_1^T L(0) B_1 + B_1^T L(0) A_1$$
$$+ B_0^T L(h) A_1 + A_1^T L(h)^T B_0 + B_1^T L(h)^T A_0 + A_0^T L(h) B_1 = + V. \quad (5.236)$$

Equation (5.219) yields

$$A_0^T L(s) - A_1^T L(h-s)^T + B_0^T L(s) + B_1^T L(h-s)^T = 0, \quad s \in [0, h]. \quad (5.237)$$

The boundary condition (5.236) may be transformed into an equivalent simpler, though slightly non-conventional, form. Putting $s = 0$ and $s = h$ in (5.237) and using (5.236), we obtain after some algebraic manipulation

$$A_0^T Q A_0 - A_1^T Q A_1 = V \quad (5.238)$$

where

$$Q = -\dot{L}(0) - \dot{L}(0)^T. \quad (5.239)$$

The matrix algebraic equation (5.238) has a unique symmetric solution Q due to our stability assumptions (5.224).

The Lyapunov system of equations (5.236), (5.237) (or, equivalently, (5.237) and (5.238)) with condition (5.220) can be solved by the Castelan–Infante approach described in section 5.3. If we define

$$U(s) = \tfrac{1}{2}[L(s) + L(h-s)] \quad (5.240)$$

$$W(s) = \tfrac{1}{2}[L(s) - L(h-s)] \quad (5.241)$$

we can rewrite equation (5.237) in the form of two ordinary matrix linear differential equations, not solved with respect to derivatives (the argument s is omitted for the sake of simplicity)

$$A_0^T \dot{W} - A_1^T \dot{W}^T + B_0^T U + B_1^T U^T = 0 \quad (5.242)$$

$$A_0^T \dot{U} + A_1^T \dot{U}^T + B_0^T W - B_1^T W^T = 0. \quad (5.243)$$

Eliminating $\dot{W}^T$ and $\dot{U}^T$ we obtain

$$A_0^T \dot{W} A_0 - A_1^T \dot{W} A_1 + A_1^T (U^T B_0 + U B_1) + (B_0^T U + B_1^T U^T) A_0 = 0 \quad (5.244)$$

$$A_0^T \dot{U} A_0 - A_1^T \dot{U} A_1 - A_1^T (W^T B_0 - W B_1) + (B_0^T W - B_1^T W^T) A_0 = 0. \quad (5.245)$$

The stability assumption (5.224) guarantees that equations (5.244) and (5.245) can be uniquely solved with respect to derivatives. Further analysis is similar to that of section 5.3; it is based on the transformation of the set of matrix differential equations (5.242), (5.243) or, equivalently, (5.244) and (5.245) into a set of ordinary linear differential equations with constant coefficients.

Sec. 5.8] Neutral systems 119

In the case of one delay, formula (5.221) gives
$$M = A_0^T L(0) A_0 + A_1^T L(0) A_1 + A_1^T L(h)^T A_0 + A_0^T L(h) A_1. \tag{5.246}$$
From (5.232) we obtain
$$J_b(x_0, \mu, \varphi) = x_0^T \int_{-h}^{0} [A_0^T L(s+h) + A_1^T L(-s)^T] y_1(s) \, ds \tag{5.247}$$
where
$$y_1(s) = A_1 \mu(s) + B_1 \varphi(s) \tag{5.248}$$
And from (5.233)
$$J_c(\mu, \varphi) = \int_{-h}^{0} \int_{-h}^{0} y_1(s)^T \begin{Bmatrix} L(\sigma - s) & \text{for } \sigma \geqslant s \\ L(s - \sigma)^T & \text{for } s > \sigma \end{Bmatrix} y_1(\sigma) \, ds \, d\sigma$$
$$= 2 \int_{-h}^{0} \int_{\sigma}^{0} y_1(\sigma)^T L(s - \sigma) y_1(s) \, ds \, d\sigma. \tag{5.249}$$

If we now perform the integration by parts in (5.247) and (5.249) and denote
$$y_0 = A_0 x_0 + A_1 \varphi(-h) \tag{5.250}$$
$$K(s) = \dot{L}(s) A_1 - L(s) B_1 \tag{5.251}$$
$$P = L(0), \quad Q = -\dot{L}(0) - \dot{L}(0)^T \tag{5.252}$$
we obtain the following expression for the integral performance criterion
$$J(x_0, \mu, \varphi) = y_0^T P y_0 + 2 y_0^T \int_{-h}^{0} K(s+h) \varphi(s) \, ds$$
$$+ \int_{-h}^{0} \varphi(s)^T A_1^T Q A_1 \varphi(s) \, ds$$
$$- 2 \int_{-h}^{0} \int_{\sigma}^{0} \varphi(\sigma)^T [A_1^T \dot{K}(s - \sigma) + B_1^T K(s - \sigma)] \varphi(s) \, ds \, d\sigma. \tag{5.253}$$

On the basis of the Lyapunov set of equations (5.236), (5.237) with condition (5.220), a similar set of equations will be derived for the function K and matrices P and Q. Equation (5.237) can be written in the form
$$A_0^T \dot{L}(s) + B_0^T L(s) = K(h-s)^T. \tag{5.254}$$

Hence
$$A_0^T \dot{L}(s)B_1 + B_0^T L(s)B_1 = K(h-s)^T B_1 \qquad (5.255)$$
and
$$A_0^T \ddot{L}(s)A_1 + B_0^T \dot{L}(s)A_1 = -\dot{K}(h-s)^T A_1. \qquad (5.256)$$
Subtracting the former equation from the latter, we obtain
$$A_0^T \dot{K}(s) + \dot{K}(h-s)^T A_1 + B_0^T K(s) + K(h-s)^T B_1 = 0, \quad s \in [0,h]. \qquad (5.257)$$

In order to derive boundary conditions for the variable K, we substitute $s = 0$ and $s = h$ into (5.251) and (5.254). This gives four equalities. The elimination of $\dot{L}(0)$ and $\dot{L}(h)$ yields the following two relationships:
$$A_0^T K(0) - K(h)^T A_1 + A_0^T PB_1 + B_0^T PA_1 = 0 \qquad (5.258)$$
$$A_0^T K(h) - K(0)^T A_1 + A_0^T L(h)B_1 + B_0^T L(h)A_1 = 0. \qquad (5.259)$$

From (5.236) and (5.259) we easily obtain
$$A_0^T PB_0 + B_0^T PA_0 + A_1^T PB_1 + B_1^T PA_1$$
$$+ K(0)^T A_1 + A_1^T K(0) - A_0^T K(h) - K(h)^T A_0 = V. \qquad (5.260)$$

Using (5.238) we finally obtain the following Lyapunov system of equations:
$$A_0^T Q A_0 - A_1^T Q A_1 = V \qquad (5.261)$$
$$A_0^T PB_0 + B_0^T PA_0 - A_0^T K(h) - K(h)^T A_0 = A_0^T Q A_0 \qquad (5.262)$$
$$A_0^T PB_1 + B_0^T PA_1 + A_0^T K(0) - K(h)^T A_1 = 0 \qquad (5.263)$$
$$A_0^T \dot{K}(s) + \dot{K}(h-s)^T A_1 + B_0^T K(s) + K(h-s)^T B_1 = 0, \quad s \in [0,h] \qquad (5.264)$$
where $P = P^T \geqslant 0$ and $Q = Q^T$.

6

Evaluation of integrals for sampled-data systems

6.1 INTRODUCTION

Integrands involving exponential and algebraic terms also arise naturally in the analysis and design of sampled-data systems. Such systems occur whenever a discrete controller is used for the control of a continuous system. This is the situation that arises when digital computing techniques are used to realize a digital control scheme for a continuous-time system. The expressions 'digital' and 'analogue' are often used in this context, and systems arising where both 'technologies' are exploited are sometimes called 'mixed', or hybrid systems.

The z-transform and its extension, the modified z-transform, are widely used in the analysis and design of such systems, and there are may fine texts on control design for sampled-data systems, for example Jury (1958), Ogata (1987) and Kuo (1980), Åstrom and Wittenmark (1984).

In this chapter we will give sufficient introduction to the z-transform for completeness. It is not the intention to reproduce what is in these standard texts but rather to show how techniques already introduced in earlier chapters may be exploited in the analysis and control of sampled-data systems extending in part some of the published techniques.

6.2 REVIEW OF z-TRANSFORM RESULTS

Consider an open-loop system, Fig. 6.1, consisting of a linear digital device with a sequence of equispaced (sampled) values as its output, a hold device, the response of which to such a sequence is a piecewise constant output, with the constants equal to the input sequence values. The output from the hold is applied as input to a linear continuous system, with a given Laplace transform transfer function.

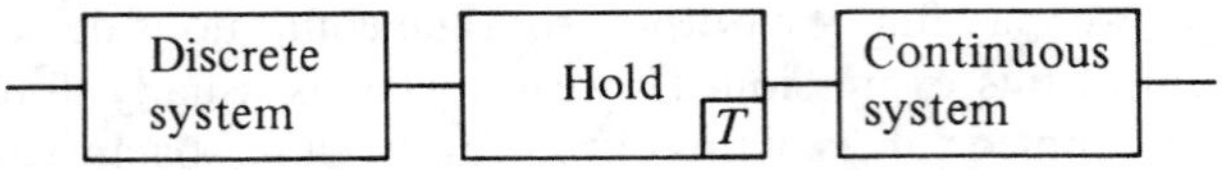

Fig. 6.1. Continuous system with discrete control.

The hold device may be considered as an interface between the discrete and the continuous. In practice a digital-to-analogue converter would be the physical device responsible for the hold operation. The T in the diagram is the time between samples, and is the duration of the hold. The name 'zero-order' hold is met for this device, which produces a constant output between sample points. It is convenient to think of the output of the discrete part as sample values of a continuous-time function $f(t)$, with the sampled values $f_r = f(rT)$, $r = 0, 1, 2, \ldots$, the elements of the input sequence to the hold device.

Consider the output from the hold, assuming that sampled values before $t = 0$ are zero. This will be

$$\sum_{r=0}^{\infty} f_r (H(t - rT) - H(t - (r+1)T))$$

with Laplace transform

$$\sum_{r=0}^{\infty} f_r \left(\frac{1}{s} \exp(-srT) - \frac{1}{s} \exp(-s(r+1)T) \right),$$

by the delay theorem of the Laplace transform. This reduces to

$$\frac{1}{s}(1 - \exp(-sT)) \sum_{r=0}^{\infty} f_r \exp(-srT).$$

Replacing $\exp(sT)$ in the summation by the variable z we have

$$\frac{1}{s}(1 - z^{-1}) \sum_{r=0}^{\infty} f_r z^{-r}.$$

The summation term which is $f_0 + f_1/z + f_2/z^2 + f_3/z^3 + \ldots$ is to be denoted by $F(z)$, which is defined as the z-transform of the sequence $\{f_0, f_1, f_2 \ldots\}$, usually shortened to $\{f_r\}$. Such expressions in $\exp(sT) = z$ which arise naturally in this way in sampled-data systems are tabulated. An early paper by Barker (1952) introduced the z-transform and contains tables, but with the constant T, which arises naturally in these expressions, replaced by unity. This practice is no longer followed. The tables in Kuo (1980) and Ogata (1987) are sufficient for most purposes.

Let the continuous system have transfer function $G(s)$; then the Laplace transform of its output will be given by

$$\mathscr{L} y(t) = \frac{1}{s}(1 - z^{-1}) F(z) G(s). \tag{6.1}$$

Hence in principle we may find the output from the continuous device by the inverse Laplace transform of this expression. However, what is usually done is to take the z-transform of the sampled values of this output, at the same spacing, and synchronous with the sampling of the discrete device. We find that the z-transform of this output sequence, $Y(z)$ say, is given by

$$Y(z) = \mathfrak{z}\left(\frac{1}{s}(1 - z^{-1})F(z)G(s)\right), \tag{6.2}$$

where $\mathfrak{z}$ denotes the z-transform of the inverse Laplace transform. This simplifies to

$$Y(z) = F(z)(1 - z^{-1})\mathfrak{z}\left(\frac{G(s)}{s}\right), \tag{6.3}$$

and the expression

$$(1 - z^{-1})\mathfrak{z}\left(\frac{G(s)}{s}\right)$$

is the z-transfer function relating the z-transform of the input sequence of the hold device to the z-transform of the sequence, obtained by sampling the output from the continuous system with Laplace transfer function $G(s)$. It is usual to denote this z-transfer function by $M(z)$ and

$$M(z) = (1 - z^{-1})\mathfrak{z}\left(\frac{G(s)}{s}\right)$$

is the interface theorem.

We denote $\mathscr{L}y(t)$ by $Y(s)$, and the z-transform of $\{y_r\}$ by $Y(z)$. The argument makes clear which transform is intended. The notation is standard.

A closed-loop sampled data control structure is shown as Fig. 6.2. Samplers are shown in the feedback path, and at the output. It is important to know that knowledge of $Y(z)$ can only provide values of $y(t)$ at the sampling times. *The z-transform only gives information at these instants.* If $y(t)$ is needed then we may use (6.1), or alternatively the modified z-transform may be used to find such inter-sample values. The modified z-transform was originally introduced by Barker for cases where the continuous plant had an explicit series delay term. If such a series delay were an integer multiple of

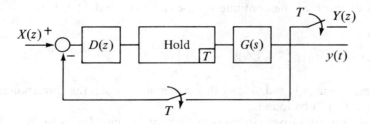

Fig. 6.2. Closed-loop sampled-data system.

T, NT, say, then there is no need of any such modification, as the rather obvious delay theorem of the z-transform allows such a delay to be represented by z^{-N}. In the absence of such a happy coincidence the extended z-transform is needed.

Add the fractional delay Td ($0 < d < 1$) to the system considered in Fig. 6.1. Then the Laplace transform of the output is

$$\mathscr{L} y(t) = Y(s) = F(z) \frac{1}{s}(1 - e^{-sT})G(s) \exp(-sTd) \tag{6.4}$$

with z replaced by $\exp(sT)$ and the z-transform of the sampled version of this output is given by

$$Y(z) = F(z)(1 - z^{-1})\mathfrak{Z}\left(\frac{G(s)\exp(-sTd)}{s}\right)$$

$$= F(z)(1 - z^{-1})\mathfrak{Z}_m\left(\frac{G(s)}{s}\right)_{m=1-d}. \tag{6.5}$$

$\mathfrak{Z}_m(G(s)/s)$ is the modified z-transform entry corresponding to the time function $\mathscr{L}^{-1}(G(s)/s)$. The modified z-transform in this example is the usual z-transform of the sequence

$$\{y\} = \{0, y(mT), y((m+1)T), y((m+2)T), \ldots\}, \tag{6.6}$$

where $m = 1 - d$. The notational change from d to m is related to the use of the modified transform for inter-sample behaviour, and is standard, if a little curious when first met.

The modified z-transform is used in two ways, either for systems where the continuous system has a series delay, or alternatively where a fictitious delay is introduced in an open-loop system, in order to find inter-sample values of delay-free system by suitable choice of m.

Delays greater than T are dealt with straightforwardly. Denote the delay by $qT + Td$, where q is an integer, and d a fraction then (6.5) is replaced by

$$Y(z) = F(z)(1 - z^{-1})z^{-q}\mathfrak{Z}_m(G(s)/s)_{m=1-d}. \tag{6.7}$$

Fig. 6.2, which shows a closed-loop system, with input sequence $\{x_r\}$, has an associated error sequence $\{e_r\}$, and a control sequence $\{u^r\}$. The continuous output is $y(t)$, and the output sequence $\{y_k\}$. Using transfer function calculations, entirely analogous to those of the continuous case, we may find $E(z)$, $U(z)$ and $Y(z)$. For example

$$U(z) = X(z)D(z)\left\{1 + D(z)(1 - z^{-1})\mathfrak{Z}\left(\frac{G(s)}{s}\right)\right\}^{-1}. \tag{6.8}$$

Replacing $F(z)$ in (6.1) and (6.2) by this expression enables the corresponding values of $\mathscr{L}(t)$, and $Y(z)$ to be found.

Knowing $E(z)$, $U(z)$ from z-transform calculations enables us to treat the system as open loop for any later calculation. This is a great simplification especially for finding inter-sample values at the output, either via the Laplace transform, or by use of the modified transform.

6.3 COST FUNCTIONALS FOR SAMPLED-DATA SYSTEMS

There are two types of integral in which one has interest in the analysis of sampled-data systems. The evaluation of

$$\sum_{r=0}^{\infty} e_r^2,$$

which via the Parseval theorem of the z-transform is given by

$$\sum_{r=0}^{\infty} e_r^2 = \frac{1}{2\pi i} \oint_{\substack{\text{unit} \\ \text{circle}}} E(z) E\left(\frac{1}{z}\right) \frac{1}{z} \, dz, \tag{6.9}$$

where $E(z)$ denotes the z-transform of $\{e_r\}$, and evaluations typified by

$$\int_0^{\infty} y^2(t) \, dt = \frac{1}{2\pi i} \int_{-i\infty}^{i\infty} Y(s) Y(-s) \, ds \tag{6.10}$$

where

$$Y(s) = E(z) D(z)(1 - \exp(-sT)) G(s)/s, \tag{6.11}$$

with $z = \exp(sT)$.

Note that the expression $E(z)$ may well be far from simple, especially if the original $G(s)$ contains delays corresponding to large values of q, i.e. the delay is large compared to the sampling interval T.

The sections that follow show relationships between the discrete and the continuous forms of Parseval's theorem, and give several examples of how the techniques for stability and evaluation of Chapter 2 and Chapter 4 may be extended for systems with and without delay, in the context of sampled-data systems.

We note that exponential terms are always present in the sampled-data case, as they arise from the hold device. Consideration of the response of the hold to the sequence $\{\exp(ir\omega T)\}$ shows that the hold device may be considered as introducing a delay of $T/2$ into any closed-loop sampled-data systems, as well as modifying the loop gain. If we compare a sampled-data system to the continuous system to which it is analogous for large sampling rates (i.e. T small), then this delay $T/2$ will be important. What the techniques of this chapter will also make possible is an exact analysis of how accurate a digital simulation might be, taking explicit account of the additional delay introduced via the inevitable hold device. If the continuous system that is modelled by a discrete simulation itself has a series delay, then the presence of the delay introduced via the hold can be taken explicitly into account. For example, a series delay h may be modelled by a shift register of N stages, where $h = (N + \frac{1}{2})T$. In simulation there will be technical limitations on the choice of T. In analytical studies of the sampled-data case we may, of course, allow T to take any convenient value, and can take limits. Clearly it is analytically helpful to investigate sampling rates which are very large, to make comparison between the discrete and the continuous. It is of interest to know how large T can be (and how small N) for the simulation still to be accurate.

6.4 STABILITY OF A SAMPLED-DATA SYSTEM

For asymptotic stability of a closed-loop sampled-data system it is necessary that its closed-loop z-transform poles lie within the unit circle, centre the origin. It is possible for a sampled-data system to be stable at the sampling instants, as checked by the z-transform, but not between samples. This is one reason why inter-sample behaviour at an output is important, and also motivates in part the evaluation of integrals such as

$$\int_0^\infty y^2(t)\,dt.$$

In this section we shall discuss a simple sampled-data system in which the continuous part has a delay which is an integer multiple, N, of the inter-sample time which we continue to denote by T. We shall explore stability as a function of the positive gain parameter K, and the value of N. Then we shall make comparisons of this system and the continuous system corresponding to the case where T is taken to be very small. Fig. 6.3 shows the system; the continuous part has a delay NT.

By the interface theorem the transfer function $M_N(z)$ relating $Y(z)$ to $E(z)$ is given by

$$(1 - z^{-1})\,\mathfrak{z}\left(\frac{G(s)}{s}\right) = (1 - z^{-1})z^{-N}\,\mathfrak{z}\left(\frac{K}{s^2}\right)$$

$$= \frac{KT}{(z-1)}z^{-N}, \quad \text{using } \mathfrak{z}\left(\frac{K}{s^2}\right) = \frac{KTz}{(z-1)^2}. \qquad (6.12)$$

The closed-loop z-transfer function is $M_N(z)/(1 + M_N(z))$, using $E(z) = X(z) - Y(z)$, so that $E(z) = X(z)/(1 + M_N(z)) = X(z)(z-1)z^N/((z-1)z^N + KT)$ and hence the closed-loop poles are the roots of

$$z^N(z-1) + KT = 0. \qquad (6.13)$$

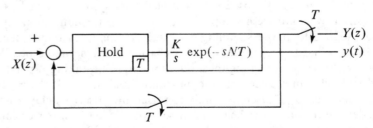

Fig. 6.3. Sampled-data system with delay NT.

Sec. 6.5] **Stability for general N** 127

At $N = 0$, there is one pole at $z = 1 - KT$, so that the system is stable for $|z| < 1$, i.e. $0 < KT < 2$. At $N = 1$, the equation satisfied by the poles is $z(z - 1) + KT = 0$. For $KT < 0.25$ the roots are real, and within the unit circle. For $0.25 < KT < 1$ the roots are complex and within the unit circle. At $KT = 1$ the roots lie on the unit circle, and the modulus of roots is greater than 1 for $KT > 1$. Hence stability for $0 < KT < 1$.

There are algebraic methods which may be followed for larger values of N, the Schur–Cohn criteria, as discussed by Jury (1958), and of course root-solving algorithms may be used with K as a parameter so that stability may be explored by a systematic search of the values of K, T and N. However, we may use a method analogous to that of the continuous case, given in Chapter 2, to be considered in the next section.

6.5 STABILITY FOR GENERAL N

The characteristic equation for the previous example for general N given by (6.13) is

$$z^N(z - 1) + KT = 0,$$

and the system is stable when all the roots of this equation lie within the unit circle. To determine the values of the parameters K, T and N when this is so it is convenient to think of N as fixed and the product KT (or equivalently K or T) as being increased from an initial starting value of zero.

At $KT = 0$ (6.13) becomes

$$z^N(z - 1) = 0. \tag{6.14}$$

The $(N + 1)$ roots of this equation consist of a root of multiplicity N at the origin and $z = 1$. For infinitesimally small positive KT the N roots initially at the origin remain close and so are within the unit circle. The root initially at $z = 1$ can also be shown to move inside the unit circle since it is given approximately by $z = 1 - KT$. Thus all $(n + 1)$ roots are within the unit circle and the system is stable.

We now follow the approach of Chapter 2 and determine for what values of the parameters there are roots of (6.13) actually on the unit circle. We first write this characteristic equation in the form

$$(z - 1) + KT z^{-N} = 0.$$

If this has roots on the unit circle, the stability boundary, then so does the equation for which z is replaced by its reciprocal. Hence $(z - 1) + KT z^{-N} = 0$ and $(1/z - 1) + KT z^N = 0$ have common, unit circle roots for values of KT satisfying $(z - 1)(1/z - 1) - (KT)^2 = 0$. This is an exact analogue of our $W = A\bar{A} - C\bar{C}$ of Chapter 2. For a root on the unit circle $z = \exp(i\theta)$ we have $(\exp(i\theta) - 1)\exp(-i\theta) - 1) - (KT)^2 = 0$, i.e. $2 - 2\cos\theta - (KT)^2 = 0$, whence $\sin(\theta/2) = KT/2$. Also $z^N = -KT/(z - 1) = -KT/(\exp(i\theta) - 1) = -KT \exp(-i\theta/2)/(2i\sin(\theta/2))$. Hence $e^{iN\theta} = -\exp(-i\theta/2)/i = \exp(-i(\theta/2 - \pi/2))$. Hence

$$N\theta = \pi/2 - \theta/2 + 2q\pi \quad \text{and} \quad \theta = (4q + 1)\pi/(2N + 1), \quad \text{for } q = 0, 1, 2, 3, \ldots.$$

$$\tag{6.15}$$

The critical value of KT is given by

$$KT = 2 \sin\left(\frac{(4q + 1)\pi}{2(2N + 1)}\right). \tag{6.16}$$

The lowest value, i.e. for $q = 0$, will give the critical value for N. In particular $KT = 2$, at $N = 0$ and at $N = 1$, $KT = 2\sin(\pi/6) = 1$, in agreement with the earlier results.

The above shows that for

$$0 < KT < 2\sin\left(\frac{\pi}{2(2N + 1)}\right)$$

all $(N + 1)$ roots of the characteristic equation lie within the unit circle and the system is therefore stable. It is also possible to show that there are no other stability regions for this example by showing that at all the positive values of KT given by (6.13) the roots do indeed leave the unit circle.

It is of interest to compare the results, at large N, with the corresponding continuous case, and this will be for the continuous system with characteristic equation $1 + (K/s)\exp(-sh) = 0$ where $h = NT + T/2$, the additional $T/2$ arising from the delay of the hold device.

Eliminating T, using $T = h/(N + \frac{1}{2})$, gives

$$Kh = 2\sin\left(\frac{\pi}{2(2N + 1)}\right)(N + \tfrac{1}{2}) \tag{6.17}$$

which it is instructive to write in the form

$$Kh = \frac{\pi}{2} \frac{\sin\left(\frac{\pi}{2(2N + 1)}\right)}{\frac{\pi}{2(2N + 1)}}. \tag{6.18}$$

The limit for large N is $\pi/2$. As this agreees with the result of Example 2.1, of Chapter 2, the analogy with the continuous case is seen to be meaningful.

Note at $N = 1$ that $Kh = 1.5$, and at $N = 3$, $Kh = (0.99163)\pi/2$, so that agreement to 0.8% is found even at low N.

This result should be treated with care, however, as the value of T in this calculation when $K = 1$, say, is given by $T = 0.44504$. This value of T was *implicit* in the calculation. The implication for simulation of the continuous case, where sample rates and hence T are given explicitly *a priori*, needs to be treated with caution.

To illustrate this, consider the sampled-data example with $N = 3$, $T = 0.1$ corresponding to a delay, in the corresponding continuous case of $h = 0.35$. The critical value of K in the continuous case is $\pi/0.7$, and in the discrete is $(2/T)\sin(\pi/14)$, giving $h_c = 4.48798$, and 4.45042 respectively with an accuracy of 0.836%.

Consider the inverse case. Let $K = 1$. Find the corresponding value of delay. In the continuous case this will be $\pi/2 = 1.5708$. In the discrete case, N will depend on the choice of T. Let $T = 0.1$ and find the corresponding N using (6.15). Now $\theta = 2\sin^{-1}(KT/2) = 2\sin^{-1}(0.05) = 0.100417$. $N = \pi/2\theta - \frac{1}{2} = 15.2014$. Hence a simulation to find N would have given stability at $N = 15$, and instability at $N = 16$. The 'theoretical value' of delay, were non-integer values possible, is $(15.2014 + 0.5) \times (0.1) = 1.57014$ and the practical value between 1.55 and 1.65 with this chosen value of 0.1 for T.

Taking a sampling rate ten times faster, i.e. $T = 0.01$, gives

$$N = \frac{\pi}{2\theta} - \frac{1}{2} = \frac{\pi}{4\sin^{-1}(0.005)} - \frac{1}{2}$$

$$= 157.0789 - \frac{1}{2} = 156.5789$$

so that $156 < N_c < 157$. This gives $T(N + \frac{1}{2})$ lying between 1.565 and 1.575. Note the large number of stages now necessary to find an accurate value of h_c.

6.6 RELATIONSHIPS BETWEEN THE DISCRETE AND THE CONTINUOUS PARSEVAL INTEGRALS

The discrete version of Parseval's theorem

$$\sum_{r=0}^{\infty} e_r^2 = \frac{1}{2\pi i} \oint_{\substack{\text{unit}\\\text{circle}}} E(z) E\left(\frac{1}{z}\right) \frac{1}{z} dz \tag{6.19}$$

is related to other integrals by means of the transformations

$$z = \frac{1+y}{1-y} \quad \text{and} \quad z = e^{i\theta}. \tag{6.20}$$

The first transformation results in the integral

$$\frac{1}{2\pi i} \int_{-i\infty}^{+i\infty} E\left(\frac{1+y}{1-y}\right) E\left(\frac{1-y}{1+y}\right) \left(\frac{1-y}{1+y}\right) \frac{2}{(1-y)^2} dy$$

$$= \frac{1}{\pi i} \int_{-i\infty}^{+i\infty} \frac{1}{(1-y)} E\left(\frac{1+y}{1-y}\right) \frac{1}{(1+y)} E\left(\frac{1+y}{1-y}\right) dy \tag{6.21}$$

which is of the form

$$\frac{1}{\pi i} \int_{-i\infty}^{+i\infty} F(y)F(-y)\,dy, \text{ the continuous form.}$$

The transformation $z = e^{i\theta}$ results in

$$\frac{1}{2\pi} \int_{-\pi}^{+\pi} E(e^{i\theta})E(e^{-i\theta})\,d\theta. \tag{6.22}$$

This integral may be further transformed by a *t*-substitution to a real integral with infinite limits.

Hence it is possible to use standard, continuous methods for integrals arising from *discrete* systems and to exploit the published tables. Åström (1970) gives iterative methods for both the discrete, and continuous cases. (6.21) contains algebraic terms only, (6.22) exponential terms only.

in *sampled-data* systems, some variables are discrete and others continuous, and in integrals involving the continuous variables, such as output, the integrands will consist of algebraic and exponential terms. It will be appropriate to extend the methods of Chapter 4 to these cases. However, it is possible to use one of the techniques of Chapter 4 in the strictly discrete case, as we have seen already in the stability example of section 6.5. We give an example of the evaluation of

$$J_D = \sum_{r=0}^{\infty} e_r^2$$

for the example of section 6.4 for the case $N = 1$, using the discrete form of Parseval's theorem. This is the method used by Sklansky (1958) for the evaluation of cost functionals in the discrete case. We shall show that the methods of Chapter 4 may be used for the case where N in an explicit z^{-N} term, arising from the presence of a shift register of N stages, is large. We shall show that the number of residues may be reduced, as in the continuous case, to a small number related to the number of poles when $N = 0$.

When N is small there is no need of such a reduction; the only difference here is that poles lie within the unit circle.

We demonstrate the Sklansky technique by using $E(z)$ of section 6.4 for the case $N = 1$, and $X(z) = z/(z-1)$.

$$E(z) = z^2/(z(z-1) + KT). \tag{6.23}$$

$$\text{Let } J_D = \sum_{r=0}^{\infty} e_r^2 = \frac{1}{2\pi i} \oint_{\text{unit circle}} E(z) E\left(\frac{1}{z}\right) \frac{1}{z}\,dz \tag{6.24}$$

$$= \frac{1}{2\pi i} \oint_{\text{unit circle}} \frac{1}{(z-\alpha_1)(z-\alpha_2)} \frac{z}{(1-z+KTz^2)}\,dz \tag{6.25}$$

where α_1, α_2 are the roots of $z^2 - z + KT = 0$. Note that as α_1, α_2 are within the unit circle, then the roots of $1 - z + KTz^2 = 0$ are external to it, and do not contribute residues to the integral. We find after summing residues that

$$J_D = \frac{(1 - (KT)^2)}{(KT)^4 - 3(KT)^2 + 2KT} = \frac{(1 - KT)(1 + KT)}{KT(2 + KT)(1 - KT)^2}$$

$$\therefore J_D = \frac{(1 + KT)}{KT(2 + KT)(1 - KT)}. \tag{6.26}$$

As a simple check, TJ_D should tend to $1/2K$ as $KT \to 0$. It does.

This example corresponded to $N = 1$ in the system of Fig. 6.3. In this case the characteristic equation was a quadratic (i.e. of order $N + 1$), and has simple roots. In simulation studies it might well be that N is large, and the standard methods of contour integration would be difficult to apply.

6.7 EXTENSION OF SKLANSKY'S METHOD FOR LARGE N

In this section we demonstrate an extension of Sklansky's method by way of a sampled-data system used for simulation of a simple continuous system with an explicit delay term. This example also demonstrates a further alternative to evaluating cost functionals, by finding the result in the discrete case and taking the limit as the sampling interval tends to zero. As before we are seeking algebraic not numerical solutions. We show that the method gives results which agree, as they should, with the earlier continuous results, and the simple example, with $N = 1$ of the previous section. When

$$E(z) = \frac{z^{N+1}}{z^N(z - 1) + KT},$$

for example, we see that a direct application of Parseval's theorem would lead to the need for an explicit expression for the $N + 1$ roots of $z^N(z - 1) + KT = 0$.

We shall avoid the difficulty associated with $N + 1$ roots by exploiting the fact that $z^{-N} = 1/z^N$ and using techniques analogous to those for the continuous systems of Chapter 4.

We shall find expressions for J_D which, when multiplied by T, should be expected to agree for large N with the continuous case.

Fig. 6.4 shows a sampled-data analogy of the continuous system of Fig. 6.4(a). We see that Fig. 6.4(b) is equivalent to that of Fig. 6.3 used in the stability example.

The variables N and T of the realization of Fig. 6.4(b) are chosen to satisfy $h = (N + \frac{1}{2})T$; the $T/2$ arises, as before, from the hold unit. The delay NT may be realized using an N-stage shift-register.

Clearly for high sample rates the sample values of error in Fig. 6.4(b) approach the corresponding values of $e(t)$ of Fig. 6.4(a) so that the approximation

$$J = T \sum_{r=0}^{\infty} e_r^2 \quad \text{to} \quad \int_0^{\infty} e^2(t)\,dt$$

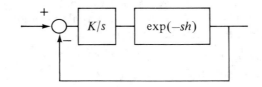

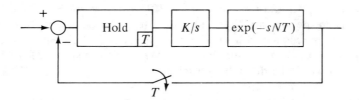

Fig. 6.4. (a) Continuous system. (b) Analogous sampled-data system.

should improve with increasing N. The sum

$$\sum_{r=0}^{\infty} e_r^2$$

is to be found by using the discrete version of Parseval's theorem. $E(z)$ is to be the z-transform of a stable sequence, and the integral is taken with the unit circle as contour.

We know from section 6.4 that a value of K exists for which $E(z)$ has all its poles within the unit circle. It follows that $E(1/z)$ has poles outside the unit circle.

From the interface theorem the open-loop transform function for the system of Fig. 6.4(b) is

$$M(z) = \frac{z-1}{z} z^{-N} \mathfrak{z}\left(\frac{K}{s^2}\right) = \frac{KT}{z^N(z-1)}. \qquad (6.27)$$

With $x(t) = H(t)$ it follows that

$$E(z) = \frac{z}{(z-1)(1+M(z))} = \frac{z^{N+1}}{z^N(z-1) + KT} \qquad (6.28)$$

and

$$E(z)E\left(\frac{1}{z}\right)\frac{1}{z} = \frac{1}{z^N(z-1) + KT} \frac{z^N}{1 - z + KTz^{N+1}}. \qquad (6.29)$$

We denote the product of these terms as $\alpha(z)\,\beta(z)$ for algebraic convenience, and note that the poles of $\alpha(z)$ are within the unit circle, and those of $\beta(z)$ external to it. Further note that when $z^N = -KT/(z-1)$ the expression $\beta(z)$ becomes $KT/[(z-1)^2 + (KT)^2 z]$.

Adopting the technique of the continuous case we write

$$\alpha(z)\beta(z) = \alpha(z)\left(\beta(z) - \frac{KT}{(z-1)^2 + (KT)^2 z}\right) + \alpha(z)\frac{KT}{(z-1)^2 + (KT)^2 z}. \quad (6.30)$$

Further we note that when $0 < KT < 2$ the roots of $(z-1)^2 + (KT)^2 z = 0$ lie on the unit circle. We denote these roots by $e^{i\theta}$ and $e^{-i\theta}$. Those roots on the unit circle are analogous to the imaginary axis roots which are found for the continuous case, and the polynomial $(z-1)^2 + (KT)^2 z = 0$, which is independent of N, is analogous to the finite polynomial in the continuous case of Chapter 4. Note the sum of roots $e^{i\theta} + e^{-i\theta} = 2 - (KT)^2$ and hence $\sin \theta/2 = KT/2$.

After substitution of expression (6.30) into expression (6.19), and indenting the contour consisting of the unit circle to exclude the poles at $e^{\pm i\theta}$, we find that the first expression of (6.30) makes no contribution, as the residues at the poles of $\alpha(z)$ are zero by construction, the poles of $\beta(z)$ are external to the contour and the roots of $(z-1)^2 + (KT)^2 z$ are outside the indented contour. If we augment the contour to include the circle at infinity as shown in Fig. 6.5 the contribution of the final term of (6.30) is given by

$$-\sum_{e^{i\theta}, e^{-i\theta}} \text{res} \frac{KT\alpha(z)}{(z - e^{i\theta})(z - e^{-i\theta})}. \quad (6.31)$$

Evaluation of expression (6.31) leads to

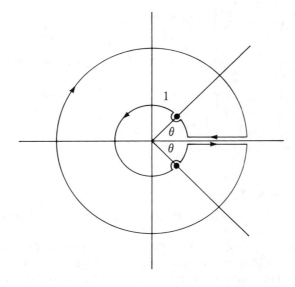

Fig. 6.5. Augmented contour.

$$J_D = \frac{1}{2K\cos\theta/2} \frac{\cos\dfrac{(2N+1)\theta}{2}}{1-\sin\dfrac{(2N+1)\theta}{2}}.\qquad(6.32)$$

Now $J \sim TJ_D$ where $T = h/(N+\tfrac{1}{2})$.

Now, from earlier,

$$\sin\frac{\theta}{2} = \frac{KT}{2} = \frac{Kh}{2N+1} \quad\text{and}\quad \theta = 2\sin^{-1}\left(\frac{Kh}{(2N+1)}\right).$$

Hence

$$J = \frac{1}{2K\cos\left(\sin^{-1}\left(\dfrac{Kh}{2N+1}\right)\right)} \frac{\cos\left\{(2N+1)\left(\sin^{-1}\left(\dfrac{Kh}{2N+1}\right)\right)\right\}}{1-\sin\left\{(2N+1)\left(\sin^{-1}\left(\dfrac{Kh}{2N+1}\right)\right)\right\}}.\qquad(6.33)$$

For large N this obviously leads to the expression

$$J = \frac{1}{2K}\frac{\cos Kh}{1-\sin Kh},$$

which agrees with the continuous case of Chapter 4.

When $N = 1$,

$$J_D = \frac{1}{2KT}\frac{\cos\left(\dfrac{3\theta}{2}\right)}{\left(1-\sin\dfrac{3\theta}{2}\right)\cos\theta/2}$$

$$= \frac{1}{2KT}\frac{(1-4\sin^2\beta)}{1-3\sin\beta+4\sin^3\beta}$$

where $\beta = \theta/2$, and

$$J_D = \frac{1}{2KT}\frac{1-(KT)^2}{1-\dfrac{3KT}{2}+\dfrac{4(KT)^3}{8}}$$

$$= \frac{(1+KT)}{KT(2+KT)(1-KT)},\qquad(6.34)$$

and $h = 3T/2$. With T replaced by $2h/3$ this agrees with (6.26) of the direct calculation for $N = 1$.

In this section we have shown that there is a discrete route to finding values for discrete cost functionals, which will lead, after taking limits, to the results for continuous cases. The method depends on exploiting the method for the continuous case, making the calculation of residues independent of N.

Sec. 6.8]   Evaluation of integrals of the mixed type arising in sampled-data systems   135

This result is useful in the case of discrete systems, being an improvement on standard results when applied to delay systems. Our motivation here is, in part, an alternative route to finding expressions for

$$\int_0^\infty e^2(t)\,dt$$

for continuous time-delay systems, and to extend Sklansky's method in a useful way, where there are explicit z^{-N} terms in expressions for $E(z)$, and N is large.

6.8 EVALUATION OF INTEGRALS OF THE MIXED TYPE ARISING IN SAMPLED-DATA SYSTEMS

As well as summing squares of the elements of a control sequence $\{e_r\}$, it is also helpful to find cost functions for continuous-time signals arising in sampled-data systems. Fig. 6.6 represents part of a sampled-data system. $G(s)$ may contain delay terms.

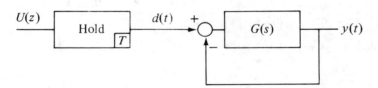

Fig. 6.6. System for Example 6.4.

For example, if $U(z)$ is given then

$$\sum_{r=0}^\infty u_r^2$$

may be found.

$$\int_0^\infty d^2(t)\,dt = T \sum_{r=0}^\infty u_r^2 \qquad (6.35)$$

where $d(t)$ denotes the output of the hold device.

$$\int_0^\infty e^2(t)\,dt = \frac{1}{2\pi i} \int_{-i\infty}^{i\infty} U(z)U\left(\frac{1}{z}\right)\frac{1-z^{-1}}{1+G(s)}\frac{1-z}{1+G(-s)}\frac{1}{(-s^2)}\,ds \qquad (6.36)$$

and

$$\int_0^\infty y^2(t)\,dt = \frac{1}{2\pi i} \int_{-i\infty}^{+i\infty} U(z)U\left(\frac{1}{z}\right) \frac{G(s)(1-z^{-1})}{1+G(s)} \frac{G(-s)(1-z)}{1+G(-s)} \frac{1}{(-s^2)}\,ds. \qquad (6.37)$$

The last pair of integrals are clearly of the mixed type with $z = \exp(+sT)$.

The complexity of the calculations may be reduced by exploiting the techniques of Chapter 4, and by other methods to be introduced in the following examples.

In the examples the following decompositions will be found helpful. Let $\beta = \exp(-aT)$, $z = \exp(sT)$, $\bar{z} = \exp(-sT)$. Then

1. $$\left(\frac{z-1}{z-\beta}\right)\left(\frac{\bar{z}-1}{\bar{z}-\beta}\right) = \frac{1}{1+\beta}\left(\frac{z-1}{z-\beta} + \frac{\bar{z}-1}{\bar{z}-\beta}\right) \qquad (6.38)$$

as may be simply verified. And

2. $$\frac{(z-1)(\bar{z}-1)}{(z-\beta)^2(\bar{z}-\beta)^2} = \frac{1}{(1+\beta)^2(1-\beta)}\left\{\frac{(z+\beta^2)(z-1)}{(z-\beta)^2} + \frac{(\bar{z}+\beta^2)(\bar{z}-1)}{(\bar{z}-\beta)^2}\right\}. \qquad (6.39)$$

This formula may be obtained by finding A, B, C by equating coefficients in the identity

$$\frac{(z-1)}{(z-\beta)^2}\frac{(\bar{z}-1)}{(\bar{z}-\beta)^2} = \frac{Az^2 + Bz + C}{(z-\beta)^2} + \frac{A\bar{z}^2 + B\bar{z} + C}{(\bar{z}-\beta)^2}.$$

Note that $z/(z-\beta)$ is the z-transform of $\{e^{-aTr}\} = \{\beta^r\}$ and

$$\frac{T\beta z}{(z-\beta)^2}$$

is the z-transform of $\{Tre^{-aTr}\}$, i.e. they are the z-transforms of sequences obtained by sampling $\exp(-at)$, and $t\exp(-at)$ respectively at $t = 0, T, 2T, 3T, \ldots$.

Example 6.1
$e(t)$ is the output from a hold with input sequence $\{\exp(-arT)\}$; find

$$\int_0^\infty e^2(t)\,dt.$$

$$E(s) = \frac{z}{z-\beta}\frac{1-z^{-1}}{s} = \frac{(z-1)}{s(z-\beta)}$$

$$\therefore J = \frac{1}{2\pi i}\int_{-i\infty}^{i\infty}\frac{(z-1)(\bar{z}-1)}{-s^2(z-\beta)(\bar{z}-\beta)}\,ds$$

Owing to the numerator terms there is no pole at the origin. Indent the contour to the right at the origin, and use the decomposition

Sec. 6.8] Evaluation of integrals of the mixed type arising in sampled-data systems

$$J = \frac{1}{2\pi i} \int \frac{1}{(1+\beta)} \frac{1}{-s^2} \left\{ \frac{z-1}{z-\beta} + \frac{\bar{z}-1}{\bar{z}-\beta} \right\} ds.$$

Close the contour to the right for the first product, which results in a zero result, and to the left for the second to obtain

$$J = \frac{-1}{(1+\beta)} \frac{d}{ds} \left\{ \frac{\bar{z}-1}{\bar{z}-\beta} \right\}_{s=0} = \frac{-1}{1+\beta} \left(\frac{-T\bar{z}}{\bar{z}-\beta} \right)_{s=0} = \frac{T}{1-\beta^2}.$$

This may be verified by finding

$$\frac{T}{2\pi i} \oint_{\substack{\text{unit} \\ \text{circle}}} E(z) E\left(\frac{1}{z}\right) \frac{1}{z} dz = \frac{T}{2\pi i} \oint_{\substack{\text{unit} \\ \text{circle}}} \frac{z}{z-\beta} \frac{\bar{z}}{\bar{z}-\beta} \frac{1}{z} dz$$

$$= \frac{T}{2\pi i} \oint_{\substack{\text{unit} \\ \text{circle}}} \frac{1}{z-\beta} \frac{1}{1-\beta z} dz = T \frac{1}{1-\beta^2},$$

from the only residue within the unit circle, that at $z = \beta = e^{-aT}$.

This example introduces a technique which is useful in more complicated cases. It would not be worth while for this example alone, owing to the easier alternative route in this case. Such easy alternatives do not always exist!

Example 6.2
As in Example 6.1 with the input replaced by $\{rT \exp(-arT)\}$. Using the second decomposition, and the z-transform quoted for the input sequence we find

$$J = \frac{1}{2\pi i} \int_{-i\infty}^{i\infty} \frac{T^2\beta^2}{(-s)^2} \frac{(z-1)(\bar{z}-1)}{(z-\beta)^2(\bar{z}-\beta)^2} ds. \quad \text{No pole at } s = 0.$$

Use decomposition 2, substituted into the integrand, indenting to the right as in the previous example, and integrating to left and right respectively to obtain

$$J = \frac{-(-T)(1+\beta^2)T^2\beta^2}{(1+\beta)^3(1-\beta)(1-\beta)^2} = \frac{T^3\beta^2(1+\beta^2)}{(1-\beta^2)^3}.$$

This time the check is more involved. It is

$$\frac{T}{2\pi i} \oint_{\substack{\text{unit} \\ \text{circle}}} \frac{Tz\beta}{(z-\beta)^2} \frac{T\bar{z}\beta}{(\bar{z}-\beta)^2} \frac{1}{z} dz$$

$$= \frac{T^3\beta^2}{2\pi i} \oint_{\substack{\text{unit} \\ \text{circle}}} \frac{z}{(z-\beta)^2(1-\beta z)^2} dz = T^3\beta^2 \frac{d}{dz} \left\{ \frac{z}{(1-\beta z)^2} \right\}_{z=\beta}$$

due to one repeated pole at $z = \beta$ inside the circle. The identical result is found from evaluating this.

Example 6.3
Find
$$J = \int_0^\infty e^2(t)\,dt$$
for the system of Fig. 6.6 when
$$U(z) = \frac{z}{z-\beta} \quad \text{and} \quad G(s) = \frac{1}{s(s+1)}.$$
Then
$$E(s) = \frac{z}{z-\beta}\frac{z-1}{z}\frac{1}{s}\left(1 + \frac{1}{s(s+1)}\right)^{-1}$$
so that
$$E(s)E(-s) = \frac{(z-1)(\bar{z}-1)(1+s)(1-s)}{(z-\beta)(\bar{z}-\beta)(s^2+s+1)(s^2-s+1)}.$$
Using decomposition 1, and symmetry
$$J = 2\sum_{(s^2+s+1)=0}\text{res}\,\frac{1}{(1+\beta)}\left(\frac{\bar{z}-1}{\bar{z}-\beta}\right)\left(\frac{1-s^2}{(s^2-s+1)(s^2+s+1)}\right).$$
Denote the roots of $s^2 + s + 1 = 0$ by $\delta_1 = -\tfrac{1}{2} + i\sqrt{3}/2$, and its conjugate δ_2. Note that $\delta_1\delta_2 = 1$, product of roots. Then
$$J = \frac{2}{1+\beta}\left\{\frac{\bar{z}_1 - 1}{\bar{z}_1 - \beta}\frac{(1-\delta_1^2)}{(\delta_1 - \delta_2)(-2\delta_1)} + \text{complex conjugate} \right\}$$
of this.
where $\bar{z}_1 = \exp(-\delta_1 T)$. Now
$$\frac{(1-\delta_1^2)}{-2\delta_1(\delta_1 - \delta_2)} = \frac{(1-\delta_1^2)\delta_2}{-2\delta_1\,\delta_2(\delta_1 - \delta_2)} = \frac{1}{2},$$
and
$$J = \frac{1}{1+\beta}\left\{\left(\frac{\bar{z}_1 - 1}{\bar{z}_1 - \beta}\right) + \text{complex conjugate}\right\},$$
$$= \frac{2\,\text{Re}}{1+\beta}\left\{\frac{(\bar{z}_1 - 1)(\bar{z}_2 - \beta)}{(\bar{z}_1 - \beta)(\bar{z}_2 - \beta)}\right\}\quad \bar{z}_2 = \exp(-\delta_2 T)$$
$$= \frac{1}{(1+\beta)}\left\{\frac{2\bar{z}_1\bar{z}_2 + 2\beta - (1+\beta)(\bar{z}_1 + \bar{z}_2)}{\bar{z}_1\bar{z}_2 + \beta^2 - \beta(\bar{z}_1 + \bar{z}_2)}\right\}$$
Now $\bar{z}_1 + \bar{z}_2 = 2\exp(T/2)\cos(\sqrt{3}T/2)$, and

Sec. 6.8]   Evaluation of integrals of the mixed type arising in sampled-data systems   139

$$J = \frac{2}{(1+\beta)} \left\{ \frac{1 + \beta \exp(-T) - (1+\beta)\exp(-T/2)\cos(\sqrt{3}T/2)}{1 + \beta^2 \exp(-T) - 2\beta \exp(-T/2)\cos(\sqrt{3}T/2)} \right\}.$$

Now $\beta = 1$ corresponds to a step output from the hold, and $J = 1$, as it should. $\beta = 0$ corresponds to $U(z) = 1$, and $J = 2(1 - \exp(-T/2)\cos(\sqrt{3}T/2))$, as it should on substituting $\beta = 0$ into the expression.

If $G(s)$ of this example is replaced by the more general

$$\frac{K^2}{s(s + 2\gamma K)},$$

where $K = 1$, $\gamma = \tfrac{1}{2}$ corresponds to what we have just found in these results, as may be checked.

$$J = \frac{1 + 4\gamma^2}{2\gamma K(1+\beta)} \left\{ \frac{e^{2KT\gamma} + \beta - (1+\beta)e^{\gamma TK}\cos(TK\sqrt{1-\gamma^2})}{e^{KT} + \beta^2 - 2\beta e^{KT}\cos(TK\sqrt{1-\gamma^2})} \right\}$$
$$+ \frac{(1-\beta)(1-\Delta\gamma^2)}{2(1+\beta)\gamma K\sqrt{1-\gamma^2}} \frac{e^{KT\gamma}\sin(\sqrt{1-\gamma^2}KT)}{(e^{2KT\gamma} + \beta^2 - 2\beta e^{KT\gamma}\cos(TK\sqrt{1-\gamma^2}))}.$$

This checks at $\beta = 1$, and $\gamma = \tfrac{1}{2}$.

Example 6.4
This is an example of an output integral, i.e.

$$\int_0^\infty y^2(t)\, dt$$

of Fig. 6.6.

The method given in Jury (1958) for such calculations is to use the modified z-transform to obtain $Y(z, m)$ and to compute the double integral

$$\frac{1}{2\pi i} \int_0^1 \oint_{\text{unit circle}} Y(z, m) Y\left(\frac{1}{z}, m\right) \frac{1}{z}\, dz\, dm.$$

The following technique, which is based on the methods met earlier in this chapter, is offered as an alternative. Jury's method may be used as a check of course, as may explicit calculations in the time domain for sufficiently simple inputs, and low-order continuous parts. The method we give here extends, as does Jury's, to $G(s)$ containing delay terms, with the delay an integer multiple of T.

Let

$$Y(s) = \frac{(z-1)}{(z-\beta)} \frac{Ka}{s(s+a)},$$

i.e. input sequence $\exp(-raT)$, and

$$G(s)/(1 + G(s)) = \frac{Ka}{s+a}.$$

The s arises from the hold device and does not usually cancel, except in the corresponding error terms.

Hence

$$J = \int_0^\infty y^2 \, dt = \frac{1}{2\pi i} \int_{-i\infty}^{i\infty} \frac{(z-1)(\bar{z}-1)}{(z-\beta)(\bar{z}-\beta)} \frac{K^2 a^2}{(-s^2)(a+s)(a-s)} \, ds.$$

Owing to the numerator terms there are no poles at the origin. Indent the contour to the right, and use decomposition 1 to obtain

$$J = \frac{1}{2\pi i} \oint_{-i\infty}^{i\infty} \frac{1}{(1+\beta)} \left\{ \frac{z-1}{z-\beta} + \frac{\bar{z}-1}{\bar{z}-\beta} \right\} \frac{K^2 a^2}{-s^2(a^2-s^2)} \, ds$$

$$= - \sum_{s=a} \text{res}\left(\frac{z-1}{z-\beta}\right) \frac{K^2 a^2}{+s^2(s^2-a^2)(1+\beta)}$$

$$+ \sum_{s=-a, s=0} \text{res}\left(\frac{\bar{z}-1}{\bar{z}-\beta}\right) \frac{K^2 a^2}{+s^2(s^2-a^2)(1+\beta)}$$

$$= -2 \sum_{s=a} \text{res}\left(\frac{z-1}{z-\beta}\right) \frac{K^2 a^2}{+s^2(s^2-a^2)(1+\beta)}$$

$$+ \sum_{s=0} \text{res}\left(\frac{\bar{z}-1}{\bar{z}-\beta}\right) \frac{K^2 a^2}{+s^2(s^2-a^2)(1+\beta)}.$$

We have omitted integrals which are zero (no enclosed poles) and have exploited symmetry to double the first integral and simplify the last.

$$\therefore J = -2 \left(\frac{z_1-1}{z_1-\beta}\right) \frac{K^2 a^2}{+a^2(2a)(1+\beta)}$$

$$+ \frac{(-T)}{1-\beta} \frac{K^2 a^2}{-a^2(1+\beta)}$$

$$= \frac{K^2 T}{(1-\beta^2)} - \frac{K^2}{a} \left(\frac{e^{aT}-1}{e^{aT}-\beta}\right) \frac{1}{1+\beta}.$$

The reader is invited to try this example via the modified transform. The dm parts of the integration are usually straightforward, as m usually appears in numerator terms only, unless the $G(s)$ contains non-integer delays.

Example 6.5
A similar example with a second-order plant with

$$Y(s) = \frac{1}{s} \frac{(z-1)}{(z-\beta)} \frac{1}{s^2+s+1}$$

leads by a similar route but including exploiting roots as in Example 6.3 to

$$J = \frac{T}{1-\beta^2} - \frac{2(1-\beta)\exp(T/2)\sin(\sqrt{3}T/2)}{(1+\beta)\sqrt{3}\left(\exp T + \beta^3 - 2\beta\exp(T/2)\cos(\sqrt{3}T/2)\right)}.$$

Example 6.6
This example introduces a delay into the $G(s)$ of Fig. 6.6. Let

$$U(z) = \frac{z}{z-\beta} \quad \text{and} \quad G(s) = \frac{K}{s}\exp(-sT)$$

Hence

$$E(s) = \frac{z}{z-\beta}\frac{(1-\bar{z})}{s}\frac{1}{1+K\exp(-sT)/s}$$

$$= \frac{z-1}{z-\beta}\frac{1}{s+K\bar{z}},$$

where $z = \exp(sT)$. Hence

$$E(s)E(-s) = \frac{z-1}{z-\beta}\frac{\bar{z}-1}{\bar{z}-\beta}\frac{1}{s+K\bar{z}}\frac{1}{-s+Kz}.$$

Hence

$$J = \int_0^\infty e^2(t)\,dt$$

$$= \frac{1}{2\pi i}\int_{-i\infty}^{i\infty}\frac{1}{(1+\beta)}\left\{\frac{z-1}{z-\beta} + \frac{\bar{z}-1}{\bar{z}-\beta}\right\}\frac{1}{s+K\bar{z}}\frac{1}{-s+Kz}\,ds$$

which, by symmetry,

$$= \frac{1}{2\pi i}\int_{-i\infty}^{i\infty}\frac{2}{(1+\beta)}\frac{\bar{z}-1}{\bar{z}-\beta}\frac{1}{s+K\bar{z}}\frac{1}{-s+Kz}\,ds$$

$$= 2\sum_{s+K\bar{z}=0}\text{res}\,\frac{1}{(1+\beta)}\frac{\bar{z}-1}{\bar{z}-\beta}\frac{1}{-s+Kz}\frac{1}{s+K\bar{z}}$$

$$= 2\sum_{s+K\bar{z}=0}\text{res}\,\frac{-1}{1+\beta}\frac{s+K}{(s+K\beta)}\frac{s}{s^2+K^2}\frac{1}{s+K\bar{z}}$$

using the substitution $\bar{z} = -s/K$.

$$\therefore J = +2\sum_{\substack{s=-K\beta \\ s=\pm iK}}\text{res}\,\frac{(s+K)s}{(1+\beta)(s+K\bar{z})(s^2+K^2)(s+K\beta)}$$

$$= \frac{2(-K\beta)(K - K\beta)}{(1 + \beta)(-K\beta + K\exp(TK\beta))(K^2 + K^2\beta^2)}$$

$$+ \frac{2}{1 + \beta}\left\{\frac{s + K}{s + K\beta}\frac{1}{s + Kz}\frac{1}{2}\right\} \text{ evaluated at } s = iK$$
$$+ \text{ complex conjugate of this}$$

$$= \frac{2\beta(\beta - 1)}{K(1 + \beta)(1 + \beta^2)(-\beta + \exp(KT\beta))}$$

$$+ 2\,\text{Re}\,\frac{(iK + K)}{(1 + \beta)(iK + K\beta)(iK + K\exp(-iKT))}.$$

The second term is

$$2\,\text{Re}\,\frac{(i + 1)(-i + \beta)(-i + \cos KT + i\sin KT)}{K(1 + \beta)(1 + \beta^2)2(1 - \sin KT)}$$

$$= \frac{(1 + \beta)\cos KT + (\beta - 1)(1 - \sin KT)}{K(1 + \beta)(1 + \beta^2)(1 - \sin KT)}.$$

Collecting terms

$$J = \frac{\cos KT}{K(1 + \beta^2)(1 - \sin KT)} + \frac{(\beta - 1)}{K(1 + \beta)(1 + \beta^2)}$$

$$+ \frac{2\beta(\beta - 1)}{K(1 + \beta^2)(-\beta + \exp(TK\beta))(1 + \beta)}$$

$$= \frac{\cos KT}{K(1 + \beta^2)(1 - \sin KT)} + \frac{(\beta - 1)}{K(\beta + 1)}\frac{(\exp(TK\beta) + \beta)}{(1 + \beta^2)(\exp(TK\beta) - \beta)}$$

At $\beta = 1$,

$$J = \frac{\cos KT}{2K(1 - \sin KT)},$$

as expected.

7
All-pass systems

7.1 INTRODUCTION

If a transfer function $F(s)$ has the property that

$$F(s)F(-s) = 1$$

for all s it follows that its gain function $|F(i\omega)|$ is unity and, in the steady state, input sinusoids of any frequency will be transmitted unchanged in amplitude. We shall call such functions, $F(s)$, all-pass functions. Let $F(0) = 1$.

If $\tilde{E}(s)$ and $\tilde{O}(s)$ are even and odd *polynomials* respectively it follows that

$$F(s) = \frac{\tilde{E}(s) - \tilde{O}(s)}{\tilde{E}(s) + \tilde{O}(s)}$$

will have this all-pass property. Alternatively it will be true if $\tilde{E}(s)$ and $\tilde{O}(s)$ are functions. We shall be concerned exclusively with stable all-pass functions, i.e. those for which the roots of $\tilde{E}(s) + \tilde{O}(s) = 0$ lie in the open LH plane. Note that $\exp(-sh)$ may be written as

$$\frac{\cosh(sh/2) - \sinh(sh/2)}{\cosh(sh/2) + \sinh(sh/2)} \tag{7.2}$$

so that in the pure delay case there is a similar decomposition to

$$\tilde{E}(s) = \cosh(sh/2), \quad \tilde{O}(s) = \sinh(sh/2).$$

We shall refer to *systems* with transfer functions in which such $F(s)$ terms appear as all-pass *systems*. Time-delay systems clearly belong to this class, as do many others. It is natural to exclude the unrealizable all-pass functions such as $F(s) = \exp(+sh)$, and this we do.

Recall that the polynomial $A(s)A(-s) - C(s)C(-s)$ arises, both in stability studies and in the evaluation of cost functionals, by the elimination of the exponential term using the fact that $\exp(-sh)\exp(+sh) = 1$, i.e. the all-pass property. This suggests that there will be strong conneections between results for delay systems where

$$E(s) = \frac{B(s) + D(s)\exp(-sh)}{A(s) + C(s)\exp(-sh)} \tag{7.3a}$$

and systems where

$$E(s) = \frac{B + DF}{A + CF} \tag{7.3b}$$

when F is an all-pass function. As elsewhere we shall assume that the polynomial A is of higher order than B, C or D, as in Chapters 2 and 4.

These connections are discussed in detail in sections 7.2 and 7.3.

In the stability section, 7.2, we shall compare stability results for systems for which the characteristic equation is

$$A(s) + C(s)F(s) = 0 \tag{7.4}$$

for given $A(s)$ and $C(s)$, and for various $F(s)$.

We shall derive, in section 7.3, expressions for the evaluation of cost functionals where the delay term is replaced by a stable all-pass function. This will serve two purposes. It will enable comparisons to be made between an exact calculation of cost, and that where the delay term is replaced by a ratio of finite polynomials, and it will serve as a method where such all-pass functions arise naturally in a transfer function.

In section 7.4 we derive relationships between what are defined in that section as forward and feedback cost functionals. This enables one integral to be derived simply from the other. The result for delay systems is derived here, exploiting properties of the delay, and by a necessarily different route it is generalized to the corresponding all-pass system.

In section 7.5, similar results are presented when the input function is a decaying exponential function.

Clearly

$$\frac{1 - sh/2}{1 + sh/2} \quad \text{and} \quad \frac{1 - sh/2 + s^2h^2/12}{1 + sh/2 + s^2h^2/12}$$

are both all-pass functions. These expressions are in fact diagonal entries in a table of Padé approximants to $\exp(-sh)$.

Padé approximants are a very important class of all-pass functions and are considered in detail in section 7.5.

In the final sections, examples are given of stability and cost functional evaluation for systems for which the all-pass functions are Padé approximants. These results are compared with those for the corresponding time-delay systems. These examples demonstrate the similarities and very important differences between the behaviour of systems which differ only in the choice of all-pass element.

7.2 STABILITY OF ALL-PASS SYSTEMS

In our discussion of stability in Chapter 2 we had in mind stability with respect to the delay h. There it was shown that the stability problem for systems with characteristic equation $A(s) + C(s)\exp(-sh) = 0$ could be solved by the following method:

(1) Check stability at $h = 0$, by delay-free methods.
(2) Check stability for h very small.
(3) Find the positive roots ω^2 of the polynomial $W(\omega^2) = A(i\omega)A(-i\omega) - C(i\omega)C(-i\omega)$, and the positive values of h which satisfy $A(s) + C(s)\exp(-sh) = 0$, at these ω values.
(4) Use sgn Re ds/dh = sgn $dW/d\omega^2$ at these values of ω to determine which values of h are stabilizing, and which are destabilizing, and hence the stability window(s).

In the context of stability of all-pass systems we consider stability with respect to a parameter appearing in the all-pass transfer function. We assume that this parameter, still conveniently called h, only appears multiplicatively with s, and vice versa, so that F in equation (7.4) is a function of the product sh. For notational convenience we shall continue to write $F(s)$ here for the all-pass transfer function.

Recall that the polynomial W arose from eliminating $\exp(-sh)$ from the equations $A(s) + C(s)\exp(-sh) = 0$ and $A(-s) + C(-s)\exp(+sh) = 0$, which are simultaneously satisfied for common roots on the imaginary axis.

If $\exp(-sh)$ is replaced by a stable all-pass function $F(s)$, then the same polynomial W will result as $F(s)F(-s) = 1$. Stability at small h is ensured owing to $F(s)$ having all poles in the open LH plane, i.e. $F(s)$ is a stable all-pass function. With these preliminaries we see that Steps (1), (2) and (3) are the same after replacement of $\exp(-sh)$ by $F(s)$.

The sgn relationship (4) is also satisfied in the all-pass case as we now show. Firstly

$$\frac{dW}{d\omega^2} = \frac{dW}{ds}\frac{ds}{d\omega^2} = \frac{i}{2\omega}\frac{dW}{ds}. \tag{7.5}$$

Denote by A' the derivative of A with respect to its argument, and similarly for the other functions. And denote $A(-s)$ by $\bar{A}$, etc.

Using (7.5) we have

$$\frac{dw}{d\omega^2} = \frac{i}{2\omega}(A'\bar{A} - \bar{A}'A - C'\bar{C} + \overline{CC'}).$$

Using $A\bar{A} = C\bar{C}$ at $s = i\omega$ satisfying $A\bar{A} - C\bar{C} = 0$ we find

$$\operatorname{sgn}\frac{dW}{d\omega^2} = \operatorname{sgn}\frac{i}{2\omega}\left\{\left(\frac{A'}{A} - \frac{C'}{C}\right) - \left(\frac{\bar{A}'}{\bar{A}} - \frac{\bar{C}'}{\bar{C}}\right)\right\}_{s=i\omega}$$

$$= \operatorname{sgn}\left\{\frac{i}{2\omega} 2i\operatorname{Im}\left(\frac{A'}{A} - \frac{C'}{C}\right)\right\} = \operatorname{sgn}\left\{-\omega\operatorname{Im}\left(\frac{A'}{A} - \frac{C'}{C}\right)\right\}. \tag{7.6}$$

This result being independent of h repeats that of Chapter 2.

Then at values of h, s satisfying $A + CF = 0$,

$$\frac{ds}{dh} = \frac{-\frac{\partial}{\partial h}(A + CF)}{\frac{\partial}{\partial s}(A + CF)} = \frac{-C\frac{\partial F}{\partial h}}{A' + C'F + CF'}. \tag{7.7}$$

We recall that the parameter h appears in the all-pass function F only in the product sh and hence $\partial F/\partial h = sF'/h$ and (7.7) becomes

$$\frac{ds}{dh} = \frac{-CsF'}{h(A' + C'F + CF')} = \frac{+sF'/F}{h(A'/A - C'/C - F'/F)} \tag{7.8}$$

on dividing numerator and denominator by $A = -CF$. Now

$$\frac{F'}{F} = \left(\frac{\tilde{E} - \tilde{O}}{\tilde{E} + \tilde{O}}\right)' \frac{\tilde{E} + \tilde{O}}{\tilde{E} - \tilde{O}} = -\frac{2(\tilde{E}\tilde{O}' - \tilde{O}\tilde{E}')}{(\tilde{E}^2 - \tilde{O}^2)}. \tag{7.9}$$

This is even in s and hence real when $s = i\omega$. Hence

$$\mathrm{Re}\,\frac{ds}{dh} = \mathrm{Re}\,\frac{i\omega\, F'/F}{h\left(\frac{A'}{A} - \frac{C'}{C} - \frac{F'}{F}\right)}\bigg|_{s=i\omega}$$

$$= \frac{\mathrm{Re}\,i\omega\frac{F'}{F}\left\{\left(-\frac{F'}{F} + \mathrm{Re}\left(\frac{A'}{A} - \frac{C'}{C}\right)\right) - i\,\mathrm{Im}\left(\frac{A'}{A} - \frac{C'}{C}\right)\right\}}{h\left|\frac{A'}{A} - \frac{C'}{C} - \frac{F'}{F}\right|^2}$$

and

$$\mathrm{sgn}\,\mathrm{Re}\,\frac{ds}{dh}\bigg|_{s=i\omega} = \mathrm{sgn}\left[\left(\frac{F'}{F}\right)_{s=i\omega} \omega\,\mathrm{Im}\left(\frac{A'}{A} - \frac{C'}{C}\right)_{s=i\omega}\right]$$

Now

$$\mathrm{sgn}\left(\frac{F'}{F}\right) = \mathrm{sgn}\left[-2\frac{\tilde{E}\tilde{O}' - \tilde{O}\tilde{E}'}{\tilde{E}^2 - \tilde{O}^2}\bigg|_{s=i\omega}\right]$$

$$= -\mathrm{sgn}(\tilde{E}\tilde{O}' - \tilde{O}\tilde{E}')_{s=i\omega}. \tag{7.10}$$

Now $\tilde{E}(s) + \tilde{O}(s) = P(s)$, say, has its roots in the open LH plane and hence, when $P(s)$ is a (finite) stable polynomial

$$P(s) = a_n \prod_{r=1}^{n}(s + p_r),$$

where p_r has positive real part, and n is finite. See Chapter 2 for the corresponding result for $F(s) = \exp(-sh)$.

Now $2\tilde{E}(s) = P(s) + P(-s)$, $2\tilde{O}(s) = P(s) - P(-s)$ and we have

Sec. 7.2]   **Stability of all-pass systems**   147

$$\tilde{E}\tilde{O}' - \tilde{O}\tilde{E}' = \frac{P\bar{P}}{2}\left\{\frac{P'}{P} + \frac{\bar{P}'}{\bar{P}}\right\}$$

and at $s = i\omega$ this is equal to

$$\frac{P(i\omega)P(-i\omega)}{2}\left\{\sum_{r=1}^{n}\frac{1}{i\omega + p_r} + \sum_{r=1}^{n}\frac{1}{-i\omega + \bar{p}_r}\right\}$$

$$\operatorname{sgn}(\tilde{E}\tilde{O}' - \tilde{O}\tilde{E}') = \operatorname{sgn}|P(i\omega)|^2 \sum_{i=1}^{n}\frac{p_r + \bar{p}_r}{\omega^2 + p_r\bar{p}_r}, \qquad (7.11)$$

i.e. $\tilde{E}\tilde{O}' - \tilde{O}\tilde{E}'$ is both real and positive. Hence

$$\operatorname{sgn}\operatorname{Re}\frac{ds}{dh} = -\operatorname{sgn}\omega\operatorname{Im}\left(\frac{A'}{A} - \frac{C'}{C}\right)$$

and hence

$$\operatorname{sgn}\operatorname{Re}\frac{ds}{dh} = \operatorname{sgn}\frac{dW}{d\omega^2}, \qquad (7.11a)$$

at values of h satisfying $A + CF = 0$, and values of ω satisfying $A\bar{A} - C\bar{C} = 0$.

Hence the slope *rule* for the stable all-pass polynomial case is the same as that for delay proved in Chapter 2.

There is a simple extension to the method of finding h values satisfying $A + CF = 0$, which makes use of the expression in F in terms of odd and even functions.

As

$$A + CF = 0, \quad -A/C = \frac{\tilde{E} - \tilde{O}}{\tilde{E} + \tilde{O}},$$

from which it is deduced that at $s = i\omega$ satisfying $A\bar{A} - C\bar{C} = 0$ (and recalling $E(i\omega)$ is real)

$$\frac{\operatorname{Im}\tilde{O}}{\tilde{E}} = \frac{\operatorname{Im}A/C}{1 - \operatorname{Re}A/C} = \frac{1 + \operatorname{Re}A/C}{\operatorname{Im}A/C}, \qquad (7.12)$$

using the result $|A/C|$ is unity, at these values of ω.

In the examples given here we find the stability window(s) for given $A(s)$ and $C(s)$, for different $F(s)$, showing how the stability behaviour is modified by the choice of all-pass function.

Example 7.1
Consider the case $A(s) = s$, $C(s) = k$ and $F(s) = (1 - sh/2)/(1 + sh/2)$ so that $\tilde{E}(s) = 1$, $\tilde{O}(s) = sh/2$. Then $W(\omega^2) = \omega^2 - k^2$, and the equation to solve for h has $\operatorname{Im}(A/C) = 1$, $\operatorname{Re}(A/C) = 0$, $\tilde{E}(i\omega) = 1$, $\tilde{O}(i\omega) = ihk/2$, so that $2/hk = 1$. $dW/d\omega^2 = 1$ at $\omega = k$. Hence one stability window $0 \leqslant hk < 2$.

For comparison recall that, when $F = \exp(-sh)$, $\tilde{E}(s) = \cosh(sh/2)$ and $\tilde{O}(s) = \sinh(sh/2)$. With the same A and C, equation (7.6) is

$$\text{Im}\frac{\sinh(i\omega h/2)}{\cosh(i\omega h/2)} = 1,$$

i.e. $\tan(kh/2) = 1$, so that $kh = \pi/2$, as in Chapter 2.

Example 7.2
Let $A(s) = s(s+1)$, $C(s) = k = \sqrt{2}$ and $F(s) = (1 - sh/2)/(1 + sh/2)$, $W(\omega^2) = \omega^4 + \omega^2 - k^2$ with its positive root $\omega^2 = \frac{1}{2}(\sqrt{(1+4k^2)} - 1) = 1$. At this value of ω

$$\frac{A}{C} = \frac{i(1+i)}{\sqrt{2}} = \frac{-1+i}{\sqrt{2}}.$$

Hence (7.6) gives

$$\frac{h/2}{1} = \frac{1/\sqrt{2}}{1 + 1/\sqrt{2}} = \frac{1}{\sqrt{2}+1}$$

so that the critical value of h is

$$\frac{2}{\sqrt{2}+1} = 2(\sqrt{2}-1).$$

This is for the only positive root, so that there is one stability window,

$$0 \leq h < 2(\sqrt{2}-1).$$

Attempting to solve the stability problem by the route: Find values of h for which $s(s+1) + \sqrt{2}(1 - sh/2)/(1 + sh/2) = 0$ leads to the cubic, $s^3 h + s^2(2+h) + s(2 - \sqrt{2}h) + 2\sqrt{2} = 0$. Equating products of inner and outer coefficients (Routh–Hurwitz) gives $hs^3 + (2+h)s^2 + (2 - \sqrt{2}h)s + 2\sqrt{2} = 0$, the solution for which is $h = 2(\sqrt{2}-1)$, as expected.

Using the all-pass method enables comparisons to be made simply between cases for different functions, $F(s)$ but for A and C unchanged.

7.3 COST FUNCTIONAL EVALUATION OF AN ALL-PASS SYSTEM

In this section an expression will be derived for the cost

$$J = \int_0^\infty e^2(t)\,dt$$

for a system with error Laplace transform

$$E(s) = \frac{B(s) + D(s)F(s)}{A(s) + C(s)F(s)}. \tag{7.13}$$

As in the earlier chapters we shall drop the explicit dependence on the argument, and denote $A(-s), B(-s), \ldots, F(-s)$ by $\bar{A}, \bar{B}, \ldots, \bar{F}$ respectively. Now

$$J = \int_0^\infty e^2(t)\,dt = \frac{1}{2\pi i}\int_{-i\infty}^{+i\infty} \frac{B+DF}{A+CF}\frac{\bar{B}+\bar{D}\bar{F}}{\bar{A}+\bar{C}\bar{F}}\,ds. \tag{7.14}$$

Under the assumption that $e(t)$ is the error function of a stable system, as may be checked by finding the roots of $A + CF = 0$ directly, or by using techniques developed in this chapter, we know that the roots of $A + CF = 0$ lie in the left half-plane, and hence those of $\bar{A} + \bar{C}\bar{F} = 0$ lie in the right half-plane. Close the contour to the left, and note than when $A + CF = 0$,

$$\frac{\bar{B}+\bar{D}\bar{F}}{\bar{A}+\bar{C}\bar{F}} = \frac{A\bar{B}-C\bar{D}}{A\bar{A}-C\bar{C}} \tag{7.15}$$

where we have used $F\bar{F} = 1$. Hence

$$J = \frac{1}{2\pi i}\int_{\Gamma_L} \left\{ \frac{B+DF}{A+CF}\left(\frac{\bar{B}+\bar{D}\bar{F}}{\bar{A}+\bar{C}\bar{F}} - \frac{A\bar{B}-C\bar{D}}{A\bar{A}-C\bar{C}}\right) + \frac{B+DF}{A+CF}\frac{A\bar{B}-C\bar{D}}{A\bar{A}-C\bar{C}} \right\}ds \tag{7.16}$$

where Γ_L is the contour consisting of the imaginary axis and the left-hand semi-circle at infinity, as in the corresponding delay case (where $F = \exp(-sh)$). This assumes that there is no contribution to the integral from the infinite left-hand semi-circle, which is to be checked in examples.

As the roots of $A\bar{A} - C\bar{C} = 0$ are not roots of $A + CF = 0$ or $\bar{A} + \bar{C}\bar{F} = 0$, we may replace the contour Γ_L by one which excludes the roots of $A\bar{A} - C\bar{C} = 0$. Hence

$$J = \sum_{A+CF=0}\mathrm{res}\,\frac{B+DF}{A+CF}\frac{A\bar{B}-C\bar{D}}{A\bar{A}-C\bar{C}},$$

as the first term in (7.16) is zero. As the integral round the infinite circle is zero, we may replace summation over the roots of $A + CF$ by the same over all other residues, with a change of sign. Hence

$$J = -\sum_{A\bar{A}-C\bar{C}=0}\mathrm{res}\,\frac{B+DF}{A+CF}\frac{A\bar{B}-C\bar{D}}{A\bar{A}-C\bar{C}}. \tag{7.17}$$

We see that (7.17) is the same as (4.15) with z replaced by F.

Example 7.3
Consider the simple system of Fig. 7.1 with $x(t) = H(t)$, and $G = K/s$,

$$E = X/(1 + KF/s)$$

$$= \frac{1}{s + KF}$$

so that $A = s$, $B = 1$, $C = K$, and $D = 0$. This is a function delay example with the feed back delay replaced by the all-pass transfer function $F(s)$.

150 **All-pass systems** [Ch. 7

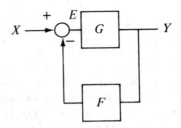

Fig. 7.1. System for Example 7.3.

$$J = -\sum_{s^2+K^2=0} \text{res} \frac{1}{s+KF} \frac{-s}{s^2+K^2}$$

$$= \sum_{s=\pm iK} \frac{1}{s+KF} \frac{s}{\frac{d}{ds}(s^2+K^2)}$$

$$= \frac{1}{2}\left(\frac{1}{iK+KF(iK)} + \frac{1}{-iK+KF(-iK)}\right)$$

$$= \frac{1}{2K} \frac{F(iK)+F(-iK)}{F(iK)F(-iK)+1-i(F(iK)-F(-iK))}$$

and hence

$$J = \frac{1}{2K} \frac{\text{Re } F(iK)}{1+\text{Im } F(iK)} \qquad (7.18)$$

after using $F(iK)F(-iK) = 1$. This formula being true for all $F(s)$, it is possible to evaluate J, with this A, B, C, D, by simple algebraic substitution, no further integrations being necessary.

When $F(s) = 1$, $J = \dfrac{1}{2K}$

When $F(s) = \exp(-sh)$, $J = \dfrac{1}{2K} \dfrac{\cos Kh}{1-\sin Kh}$

When $F(s) = \dfrac{1-sh/2}{1+sh/2}$, $J = \dfrac{1}{2K} \dfrac{2+hK}{2-hK}$.

Hence the corresponding delay-free and delay results are recovered, and in the third case we infer a stability limit $hK < 2$, i.e. when J becomes infinite for the all-pass system with the all-pass transfer function $(1-sh/2)/(1+sh/2)$ in its feedback path.

Example 7.4

For the system of Fig. 7.1 with the same input $X = 1/s$, but with $G = K/(s(s+1))$, we have

$$E = \frac{(s+1)}{s(s+1) + KF} \tag{7.19}$$

with $A = s(s+1)$, $B = 1+s$, $C = K$, $D = 0$. When $D = 0$, (7.5) simplifies to

$$J = -\sum_{A\bar{A} - C\bar{C} = 0} \text{res} \frac{AB\bar{B}}{(A + CF)(A\bar{A} - C\bar{C})} \tag{7.20}$$

and

$$W = A\bar{A} - C\bar{C} = -s^2(1-s^2) - K^2 = s^4 - s^2 - K^2 = (s^2 - \alpha^2)(s^2 + \beta^2)$$

where $\alpha^2 = \frac{1}{2} + (K^2 + \frac{1}{4})^{1/2}$, $\beta^2 = (K^2 + \frac{1}{4})^{1/2} - \frac{1}{2}$. Note that $\alpha^4 - \alpha^2 = K^2$, and $\beta^4 + \beta^2 = K^2$, $\alpha^2 \beta^2 = K^2$, $\alpha^2 - \beta^2 = 1$.

$$\therefore J = -\sum_{W=0} \text{res} \frac{s(1+s)(1-s^2)}{(s(1+s) + KF)(s^4 - s^2 - K^2)}$$

$$= -\sum_{r=1}^{4} \frac{s_r(1+s_r)(1-s_r^2)}{(s_r(1+s_r) + KF_r)(4s_r^3 - 2s_r)} \tag{7.21}$$

where $s_1 = \alpha$, $s_2 = -\alpha$, $s_3 = i\beta$, $s_4 = -i\beta$, and $F_r = F(s_r)$

$$= -\frac{1}{2}\sum_{r=1}^{4} \frac{(1+s)(1-s^2)}{(2s^2 - 1)(s(1+s) + KF_r)} \frac{(-s(1-s) + K\bar{F}_r)}{(-s(1-s) + K\bar{F}_r)}\bigg|_{s=s_r}$$

to obtain a denominator which is even in s, which is

$$J = -\frac{1}{2}\sum_{r=1}^{4} \frac{-s(1-s^2)^2 + K\bar{F}(1+s)(1-s^2)}{(2s^2-1)\{-s^2(1-s^2) + K^2 + Ks^2(F+\bar{F}) - Ks(F_r - \bar{F}_r)\}}. \tag{7.22}$$

Now at each root s_r, $s^4 - s^2 = K^2$. By the symmetry of the roots, the first numerator term vanishes after summation.

$$J = -\frac{1}{2}\sum_{r=1}^{4} \frac{K\bar{F}(1+s)(1-s^2)}{(2s^2-1)\{2K^2 + Ks^2(F+\bar{F}) - Ks(F-\bar{F})\}}\bigg|_{s=s_r}$$

$$= \frac{1}{2}\sum_{r=1}^{4} \frac{(1-s^2)(1+s)\bar{F}}{(1-2s^2)(2K + s^2(F+\bar{F}) - s(F-\bar{F}))}\bigg|_{s=s_r} \tag{7.23}$$

$$= \frac{1}{2}\left\{\frac{(1-\alpha^2)}{(1-2\alpha^2)} \frac{\text{Ev } F(\alpha) - \alpha \text{ Odd } F(\alpha)}{(K + \alpha^2 \text{ Ev } F(\alpha) - 2\alpha \text{ Odd } F(\alpha))}\right.$$

$$\left. + \frac{(1+\beta^2)}{(1+2\beta^2)} \frac{\text{Re } F(i\beta) + \beta \text{ Im } F(i\beta)}{(K - \beta^2 \text{ Re } F(i\beta) + \beta \text{ Im } F(i\beta))}\right\}. \tag{7.24}$$

This is a result true for all stable $F(s)$, and the given A, B, C, D. This will be checked for $F = \exp(-sh)$. The reader is invited to check it for $F = 1$, the delay-free case.

At $F = \exp(-sh)$

$$F(\alpha) = \cosh \alpha h - \sinh \alpha h = \text{Ev } F(\alpha) + \text{Odd } F(\alpha)$$
$$F(i\beta) = \cos \beta h - i \sin \beta h = \text{Re } F(i\beta) + i \text{ Im } F(i\beta)$$
$$\therefore J = \tfrac{1}{2} \left\{ \frac{(1-\alpha^2)}{(1-2\alpha^2)} \frac{\cosh(\alpha h) + \alpha \sinh(\alpha h)}{K + \alpha^2 \cosh(\alpha h) + 2\alpha \sinh(\alpha h)} \right.$$
$$\left. + \frac{(1+\beta^2)}{(1+2\beta^2)} \frac{\cos \beta h - \beta \sin \beta h}{K - \beta^2 \cos \beta h - \beta \sin \beta h} \right\}. \tag{7.25}$$

We may apply a stability check to the second denominator: when $F = z$ the characteristic equation is $s(s+1) + K \exp(-sh) = 0$ in the delay case. At $s = i\beta$ we have $i\beta(1 + i\beta) + K(\cos \beta h - i \sin \beta h) = 0$. Take the sum of squares of the real and the imaginary part of this, and $\beta^4 - \beta^2 - K^2 = 0$, to obtain

$$(K \cos \beta h - \beta^2)^2 + (\beta - K \sin \beta h)^2$$
$$= K^2(\cos^2 \beta h + \sin^2 \beta h) - 2\beta^2 K \cos \beta h + \beta^4 + \beta^2 - 2K\beta \sin \beta h$$
$$= 2(K^2 - K\beta^2 \cos \beta h - K\beta \sin \beta h),$$

containing the denominator factor appearing in J, as it should, if correct. Having (7.24) reduces the calculation of cost for the given A, B, C, D to simple algebraic substitutions, for any all-pass $F(s)$.

7.4 AN ALL-PASS COST-DIFFERENCE THEOREM

In this section, and the one to follow, we shall investigate the difference between two cost functionals, arising from the same system. We first derive the result for the pure delay case, which will be found to be straightforward—no contour integrations being necessary. The generalization of the result to the all-pass case follows at greater length. In addition we need to exploit some simple identities of all-pass functions, i.e. $F\bar{F} = 1$, already met in the definition, and the following:

$$(1 + F)(1 + \bar{F}) = (1 + F) + (1 + \bar{F})$$
$$(1 + F)(1 - \bar{F}) = F - \bar{F}, \quad (1 - F)(1 + \bar{F}) = \bar{F} - F. \tag{7.26}$$

Consider the two systems shown in Fig. 7.2(a), 7.2(b). Let the input to both systems be $x(t) = H(t)$. Figure 7.2(a) is denoted the feedback delay case, and Fig. 7.2(b) the forward delay case. Denote the corresponding errors defined as input–output by $e_B(t)$ and $e_F(t)$ respectively, and the corresponding costs by

$$J_B = \int_0^\infty e_B^2(t)\,dt \quad \text{and} \quad J_F = \int_0^\infty e_F^2(t)\,dt.$$

We seek to find an expression for $J_B - J_F$. On the assumption of zero initial conditions we have $z(t) = y(t - h) = 0$ for $t < h$. As $x(t) = H(t)$ we have $x(t - h) = x(t)$ for $t > h$.

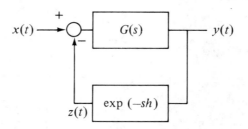

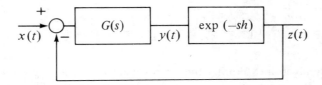

Fig. 7.2. (a) Feedback path delay system. (b) Forward path delay system.

Hence

$$J_F = \int_0^\infty (x(t) - z(t))^2 \, dt$$

$$= \int_0^h (x(t) - z(t))^2 \, dt + \int_h^\infty (x(t) - z(t))^2 \, dt$$

$$= h + \int_h^\infty (x(t-h) - y(t-h))^2 \, dt$$

$$= h + J_B.$$

Hence

$$J_B = J_F - h. \tag{7.27}$$

This result is very useful in deriving one integral from the other—it is often the case that one is much easier to calculate, owing to a simpler integrand. Examples of this are to be met in Chapter 9.

Similar motivation leads to finding the corresponding result when the delay is replaced by an all-pass function $F(s)$.

Now

$$E_F = \frac{X}{1+FG} \quad \text{and} \quad E_B = X - Y = \frac{X(1-G(1-F))}{1+FG}.$$

Hence

$$J_F - J_B = \frac{1}{2\pi i} \int_{-i\infty}^{+i\infty} X\bar{X}\left\{\frac{1}{(1+FG)(1+\bar{F}\bar{G})} - \frac{(1-G(1-F))(1-\bar{G}(1-\bar{F}))}{(1+FG)(1+\bar{F}\bar{G})}\right\}ds \tag{7.28}$$

$$= \frac{1}{2\pi i} \int_{-i\infty}^{+i\infty} X\bar{X}\left(\frac{G(1-F)}{1+FG} + \frac{\bar{G}(1-\bar{F})}{1+\bar{F}\bar{G}}\right)ds \tag{7.29}$$

after considerable algebra exploiting the simple identities in $F, \bar{F}$ given earlier.

As G has at least one pole at the origin then the integrand (7.28) has no poles at the origin even if $X(s) = 1/s$, so that the integral in (7.28) may be replaced by the contour integral closed to the left by the infinite left half semi-circle but excluding the origin. This will apply to (7.29) also. Note that the two parts of the integrand of (7.29) do not have this property when considered separately owing to the numerator G terms. However, $F(0)$ is unity so that $1 - F(s)$ tends to zero as s tends to zero, and hence there is a simple pole at $s = 0$.

Hence (7.29) may be written, for $X = 1/s$, as

$$-\frac{1}{2\pi i}\int_{\Gamma_L} \frac{1}{s^2}\left\{\frac{G(1-F)}{1+FG} + \frac{\bar{G}(1-\bar{F})}{1+\bar{F}\bar{G}}\right\}ds \tag{7.30}$$

$$= -\frac{1}{2\pi i}\int_{\Gamma_L} \frac{1}{s^2}\frac{G(1-F)}{1+FG}ds,$$

as the second integrand has no poles within the contour. Hence

$$J = -\sum_{1+FG=0}\text{res}\,\frac{1}{s^2}\frac{G(1-F)}{1+FG} = \sum_{s=0}\text{res}\,\frac{1}{s^2}\frac{G(1-F)}{1+FG} \tag{7.31}$$

$$= \frac{d}{ds}\left\{\frac{G(1-F)}{1+FG}\right\}_{s=0} = -F'(0)\lim_{s\to 0}\frac{G}{1+FG}. \tag{7.32}$$

But

$$\frac{1}{s}\left(1 - \frac{G}{1+FG}\right) = E_B \quad \text{and} \quad \lim_{s\to 0} sE_B = 0.$$

Hence

$$\lim_{s\to 0}\frac{G}{1+FG} = 1$$

and
$$J_F - J_B = -F'(0). \tag{7.33}$$

Clearly when $F(s) = 1$, $J_F = J_B$. When
$$F(s) = \frac{(1 - sh/2)}{(1 + sh/2)}, \quad F'(0) = -h$$

and we have $J_F - J_B = h$, as in the pure delay case. In fact for any $F(s)$ satisfying (7.1), and for which the leading term of $\tilde{O}(s)$ is $sh/2$, the same result, i.e. $J_F - J_B = h$, will apply.

Note also that $-(d/ds)(\exp(-sh))$ is also equal to h at $s = 0$, so that the result is also true for the pure delay case.

7.5 EXPONENTIAL INPUTS TO ALL-PASS SYSTEMS

Returning to expression (7.22) and replacing X by $1/(\alpha + s)$ and $\bar{X}$ by $1/(\alpha - s)$, and assuming α is not a pole of $G(s)$ results in

$$J_F - J_B = \frac{1}{2\pi i} \int_{-i\infty}^{+i\infty} \frac{1}{(\alpha - s)(\alpha + s)} \left\{ \frac{G(1-F)}{1+FG} + \frac{\bar{G}(1-\bar{F})}{1+\bar{F}\bar{G}} \right\} ds \tag{7.34}$$

$$= \frac{1}{\pi i} \int_{-i\infty}^{+i\infty} \frac{1}{(\alpha - s)(\alpha + s)} \frac{G(1-F)}{1+FG} ds$$

by symmetry. Closing the contour to the right, with a single pole $s = \alpha$, enclosed, gives

$$J_F - J_B = \frac{1}{\alpha} \frac{G(\alpha)(1 - F(\alpha))}{1 + F(\alpha)G(\alpha)} \tag{7.35}$$

with the limit as $\alpha \to 0$ of $-F'(0)$, as it should be.

Let us compare this with the case of an exponential input to the corresponding delay system, where $F(s)$ is replaced by $\exp(-sh)$.

From Fig. 7.2(a) with $X(t) = \exp(-\alpha t)H(t)$

$$\mathscr{L}(x(t) - y(t)) = X\left(1 - \frac{G}{1+Gz}\right) = \frac{1}{s+\alpha}\left(1 - \frac{G}{1+Gz}\right)$$

$$J_F = \int_0^\infty (x-z)^2 \, dt = \int_0^h (x-z)^2 \, dt + \int_h^\infty (x-z)^2 \, dt$$

$$= \int_0^h x^2(t) \, dt + \int_0^\infty (e^{-\alpha h}x(t) - y(t))^2 \, dt$$

$$= \int_0^h e^{-2\alpha t}\,dt + \int_0^\infty ((e^{-\alpha h} - 1)X(t) + x(t) - y(t))^2\,dt$$

$$= \frac{1}{2\alpha}(1 - e^{-2\alpha h}) + (1 - e^{-\alpha h})^2 \int_0^\infty x^2(t)\,dt + \int_0^\infty (x(t) - y(t))^2\,dt$$

$$- 2(1 - e^{-\alpha h}) \int_0^\infty x(t)(x(t) - y(t))\,dt$$

$$= +\frac{1}{2\alpha}1 - e^{-2\alpha h} + 1 - 2e^{-\alpha h} + e^{-2\alpha h}) + J_B$$

$$- 2(1 - e^{-\alpha h}) \int_0^\infty e^{-\alpha t}(x(t) - y(t))\,dt$$

$$= J_B + \frac{(1 - e^{-\alpha h})}{\alpha} - 2(1 - e^{-\alpha h})\mathscr{L}(x(t) - y(t))_{s=\alpha}$$

$$= J_B + \frac{(1 - e^{-\alpha h})}{\alpha}\left\{1 - \left(1 - \frac{G}{1 + Gz}\right)_{s=\alpha}\right\}$$

$$= J_B + \frac{(1 - e^{-\alpha h})}{\alpha} \frac{G(\alpha)}{1 + G(\alpha)e^{-\alpha h}}. \tag{7.36}$$

This agrees with (7.28) on replacing $F(\alpha)$ with $\exp(-sh)_{s=\alpha}$. Hence the delay result and the all-pass result agree with alternative methods of derivation, the second depending on the explicit use of delay.

7.6 PADÉ APPROXIMANTS OF THE EXPONENTIAL FUNCTION

A Padé approximant to a function $f(x)$ is a ratio of two finite-order polynomials in x, whose coefficients are chosen so that the Maclauren series representation of the Padé approximant agrees with the first N terms of the corresponding series for the given function, such that N is as large as possible. If the polynomials are of order L, M respectively, the $L + M + 1$ free coefficients may be chosen so that N is at least $L + M + 1$.

A table of such polynomials corresponding to choices of L, M is called the Padé table for the function. We shall be concerned, mostly, with the diagonal entries of such a table, that for the exponential function, $\exp(-sh)$. It can be shown (Padé, 1892, 1899, Wall, 1948 Ch. 20, Baker and Graves-Morris, 1981) that the diagonal entries of the Padé table for $\exp(-sh)$ are given by

$$\exp(-sh) \approx P^{(M,M)}(sh) = \frac{\sum_{j=0}^{M} \frac{[M]^j}{[2M]^j}(-sh)^j}{\sum_{j=0}^{M} \frac{[M]^j}{[2M]^j}(sh)^j} \tag{7.37}$$

where $[M]^j = M(M-1)\ldots(M-j+1)$, i.e. the j leading terms of $M!$. The first three entries of the table are

$$P^{(1,1)}(sh) = \frac{1 - \frac{sh}{2}}{1 + \frac{sh}{2}}, \quad P^{(2,2)}(sh) = \frac{1 - \frac{sh}{2} + \frac{h^2 s^2}{12}}{1 + \frac{sh}{2} + \frac{h^2 s^2}{12}},$$

$$P^{(3,3)}(sh) = \frac{1 - \frac{sh}{2} + \frac{h^2 s^2}{10} - \frac{h^3 s^3}{120}}{1 + \frac{sh}{2} + \frac{h^2 s^2}{10} + \frac{h^3 s^3}{120}}.$$

These, and all the other diagonal entries given by (7.30), are clearly all-pass functions in s. The use of the Padé table for the exponential, and other functions in control, and simulation studies are given by Storer (1957), Fifer (1961, Vol. 4), and Truxal (1958). Use of the Padé approximants for sine, cosine and hyperbolic functions are given in detail by Kogbetliantz, in Ralston and Wilf (1965, Chapter 1), where the use of such functions in numerical analysis is also discussed.

Choksy (1960) and Marshall (1979) have drawn attention to the pitfalls of replacing the exponential series by a truncated version especially in stability studies. What it is now possible to do, in the light of Chapters 2 and 4, and the earlier sections of this chapter, is to compare the effects on stability and cost functional evaluation brought about by the use of a Padé approximant used as a replacement of the exponential function. It is reasonable to conjecture that the use of the diagonal elements of the Padé table is likely to produce better results (in some sense) than those of a truncated series, which being off-diagonal elements of the Padé table do not have the important all-pass property.

7.7 PADÉ APPROXIMANT AND STABILITY

It has been shown in section 7.2 that there are close relationships in stability analysis for delay systems and all-pass systems, the expression $W(\omega^2) = A\bar{A} - C\bar{C}$ being common to both methods. However, the behaviour of a system for which the delay is replaced by an all-pass function may be qualitatively different, as will be shown in the examples that follow.

Even if an exponential transfer function is replaced by an all-pass Padé approximant, i.e. leading terms of the Maclaurin series agree, it will not necessarily be the case that even the first stability window will be accurately found. In examples there is usually some qualitative agreement *for small h*, but behaviour may be very different at large h, even when the Padé approximant is a diagonal one with unity gain at all frequencies. The qualitative behaviour improves, as might reasonably be conjectured, with increased order of Padé approximant—but at greater algebraic complexity.

Example 7.5
The all-pass system that we shall consider is that for which $A(s) = s^2 + s + 2$ and $C(s) = \sqrt{2}$, and where $B(s) = 1$. $D(s) = 0$. We may apply expression (7.12) to this case, for $F(s) = \exp(-sh)$, and then repeat the analysis for the first three Padé approximants to $\exp(-sh)$. $W(\omega^2) = A\bar{A} - C\bar{C} = (\omega^2 - 1)(\omega^2 - 2)$ in all of these cases.

For the pure delay

$$\text{Im } \tilde{O}(i\omega) = \sin \omega h/2$$

$$\tilde{E}(i\omega) = \cos \omega h/2$$

and using (7.12) we have

$$\tan\left(\frac{\omega h}{2}\right) = \frac{\text{Im}(A/C)}{1 - \text{Re}(A/C)},$$

where $\text{Im}(A/C)$, $\text{Re}(A/C)$ have to be found at $\omega = 1$ (stabilizing), and $\omega = \sqrt{2}$ (destabilizing). The right-hand sides of this expression will also be unchanged when the delay is replaced by a Padé all-pass function.

$$\text{Re}(A/C) = \frac{2 - \omega^2}{\sqrt{2}}; \quad \text{Im}(A/C) = \omega/\sqrt{2}$$

so that the RHS is $\sqrt{2} + 1$ when $\omega = 1$, and unity when $\omega = \sqrt{2}$. At $\omega = 1$ the (stabilizing) values of h are $h = (3\pi)/4 + 2q\pi$, $q = 0, 1, 2, \ldots$, and at $\omega = \sqrt{2}$ the destabilizing values are $\pi\sqrt{2}(\frac{1}{4} + q)$, $q = 0, 1, 2, \ldots$. On inspection of these values it is found that there are three stability windows in the delay case (see Table 7.1).

Table 7.1. Critical values of delay

$\omega_1 = 1$	Stabilizing	—	0.75π	—	2.75π	—	—
$\omega_2 = \sqrt{2}$	Destabilizing	0.35π	—	1.76π	—	3.18π	4.56π

$$0 \leqslant h < \sqrt{2}\,\pi/4, \quad 3\pi/4 < h < 5\sqrt{2}\,\pi/4 \quad \text{and} \quad 11\pi/4 < h < \frac{9\pi\sqrt{2}}{4}.$$

This calculation is the special case of Example 2.8 for $b = \sqrt{2}$. The number of stability windows (3) is as predicted. We now find the windows for the first three Padé approximants to $\exp(-sh)$. The destabilizing values are still at $\omega = \sqrt{2}$, and the stabilizing ones at $\omega = 1$. For the first Padé approximant $\tilde{E}(i\omega) = 1$, $\text{Im } \tilde{O}(i\omega) = (\omega h)/2$

so that we find a destabilizing $h = \sqrt{2}$, and one stabilizing value of $2(1 + \sqrt{2})$. Clearly there is stability for $h = 0$ (for all cases), so that for the first Padé there are two windows

$$0 \leq h < \sqrt{2} \quad \text{and} \quad 2(1 + \sqrt{2}) < h < +\infty.$$

For the second all-pass Padé approximant

$$\tilde{E}(i\omega) = 1 - \frac{h^2 \omega^2}{12}, \quad \text{Im } \tilde{O}(i\omega) = \omega h/2.$$

At $\omega = \sqrt{2}$ the destabilizing values of h are given by

$$\frac{\sqrt{2}h}{2} \bigg/ \left(1 - \frac{2h^2}{12}\right) = 1, \quad \text{yielding} \quad \frac{h^2}{6} + \frac{h}{\sqrt{2}} - 1 = 0,$$

with a positive root of $3((7/6)^{1/2} - 1/\sqrt{2}) = 1.12$, and a negative root which is of no interest. At $\omega = 1$ there is a stabilizing value of $h = 2.44$.

Hence for the second Padé approximant there are two windows

$$0 \leq h < 1.12 \quad \text{and} \quad 2.44 < h < +\infty.$$

The third-order Padé for which

$$\tilde{E}(i\omega) = 1 - \frac{\omega^2 h^2}{12} \quad \text{and} \quad \text{Im } \tilde{O}(i\omega) = \frac{\omega h}{2} - \frac{\omega^3 h^3}{120}$$

yields the cubic equations

$$\frac{\sqrt{3}h^3}{60} - \frac{h^2}{5} - \frac{h}{\sqrt{2}} + 1 = 0$$

for the destabilizing values at $\omega = \sqrt{2}$ and

$$\frac{h^3}{120} + (1 + \sqrt{2})(1 - h^2/10) - h/2 = 0$$

for the stabilizing values. For this case there are three stability windows

$$0 \leq h < 1.12, \quad 2.36 < h < 9.184, \quad \text{and} \quad 30.62 < h < +\infty.$$

We see that there are still large values of h beyond which stability is found.

Figure 7.3 shows these windows graphically, together with the results for the delay case. The graphs show that there are some similarities for low h, but for large h the behaviour is qualitatively different—i.e. in the three Padé cases, stability occurs for large h. Note, the number of stability windows now depends on the order of the Padé approximant.

This example, which may be taken as typical, shows the inadvisability of using the Padé approximant as a replacement for a delay in a stability analysis. Inspection of the phase versus frequency graph of the Padé approximants shows that the first stability interval is over-estimated when Padé is used, and approaches the correct delay value as the Padé order is increased.

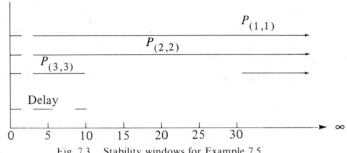

Fig. 7.3. Stability windows for Example 7.5.

7.8 PADÉ APPROXIMANT AND THE EVALUATION OF COST FUNCTIONALS

The methods of section 7.3, especially expression (7.17), enable cost functionals to be found for the Padé case, as examples of all-pass functions $F(s)$. In fact example 7.3 has already found such a cost. However, (7.18), and (7.24) would allow the cost for any diagonal Padé approximant to be written in whithout further integration; it is only necessary to find Re $F(iK)$ and Im $F(iK)$ for the given Padé expression.

Further, (7.17) enables a similar substitution to be made where $E(s)$ is given by (7.13) with F replaced by the corresponding Padé expression. The worked example that follows is chosen to show that analysis of cost for the case

$$E(s) = \frac{1}{(s+1)^2 + 2sF(s)},$$

where $F(s)$ is either a pure delay $\exp(-sh)$ or the first Padé approximant

$$\frac{1 - sh/2}{1 + sh/2}.$$

The calculation is given to allow comparisons to be made of a delay system cost, and that for a corresponding all-pass, Padé approximant, system.

Example 7.6
Consider the example of an all-pass system given by

$$E = \frac{1}{(s+1)^2 + 2sF(s)}, \tag{7.38}$$

i.e. $A = (s+1)^2$, $B = 1$, $C = 2s$, $D = 0$.

$$A\bar{A} - C\bar{C} = (s^2 + 1)^2$$

and

$$J = -\sum_{s=\pm i} \text{res} \frac{(s+1)^2}{(s^2+1)^2((s+1)^2 + 2sF(s))}$$

using (7.17).

$$\therefore J = -2 \,\text{Re}\, \frac{\partial}{\partial s} \frac{(s+1)^2}{(s+i)^2} \frac{1}{(s+1)^2 + 2sF(s)}\bigg|_{s=i}.$$

After routine algebra we obtain

$$J = \tfrac{1}{2} \text{Re} \left\{ \frac{1+i}{1+F(i)} - \frac{1+F'(i)}{(1+F(i))^2} \right\}. \tag{7.39}$$

When $F(s)$ is the delay function $\exp(-sh)$, $F'(s) = -he^{-sh}$ and

$$J_{\text{delay}} = \frac{1 + h - \sin h}{4(1 + \cos h)},$$

which suggests stability for $0 \leqslant h < \pi$.

When $F(s)$ is the first Padé all-pass

$$F(s) = \frac{1 - hs/2}{1 + hs/2}, \quad F'(s) = \frac{-h}{(1 + sh/2)^2}$$

and substitution in (7.39) results in

$$J_{\text{Padé}} = \frac{4 + h^2}{32}, \tag{7.40}$$

which implies stability for all positive h, owing to the constant denominator. This shows again that when the delay is replaced by a Padé approximant, the qualitative stability properties are not maintained.

Expansion of J_{delay} gives

$$J_{\text{delay}} = \frac{1}{8}(1 + h^2/4 + h^3/6 + h^4/24 \ldots).$$

It will be seen that this and (7.40) agree as far as the term in h^2, i.e. to the same number of terms for which there is matching of the Padé approximant, and the function $\exp(-sh)$. This is a general result.

In the calculation of cost functionals for which the value of h in the PA is finite it is of course possible to use delay-free methods: however, as is seen in this example, it is possible by exploiting the all-pass property to obtain the delay and the all-pass results by a common route. One could similarly have found J from the higher-order Padé approximants. Clearly the manipulation required, once the basic all-pass solution is known, in the form given by (7.39), is not very great, being simple complex algebra, with no further integration, or use of tables, needed.

Comparing leading coefficients serves as a check, as can the alternative delay-free calculation when $\tilde{O}(s)$ and $\tilde{E}(s)$ are finite polynomials.

8

Application to PID control

8.1 TIME-DELAY SYSTEMS WITH PID-CONTROLLERS

Closed-loop feedback systems with PID-controllers are the basic and most frequently used systems of automatic control. We call such systems conventional, as opposed to the more sophisticated ones which include computers producing control signals according to various elaborate algorithms. A typical one-loop conventional control system is shown in Fig. 8.1. The PID-controller involves action of three kinds: proportional (P), integral(I) and differential (D). It is described by the relationship between its input ε and output u:

$$u(t) = a_R \varepsilon(t) + b_R \left[\int_0^t \varepsilon(\tau) \, d\tau + \eta \right] + c_R \dot{\varepsilon}(t), \quad t \geq 0 \tag{8.1}$$

where a_R, b_R and c_R are non-negative constants. The additive term η accounts for initial condition in the integration part of the controller. We assume that the differential action is perfect, which may be not precisely satisfied in real systems. The P-, I-, PD- and PI-controllers are special cases of the PID-controller; the respective formulae for control signals u are obtained from the general formula (8.1) by setting the appropriate coefficients a_R, b_R or c_R equal to zero. The remaining coefficients are adjustable. Their values are chosen so that some desired features of the system might be obtained. Asymptotic stability is the basic requirement; others are expressed in terms of time response to special inputs or initial conditions, or in terms of frequency characteristics. The most frequently used special inputs are step functions, approximations of delta functions, sinusoidal inputs and stochastic noise. Integral performance criteria of the form considered in earlier chapters are convenient and commonly used

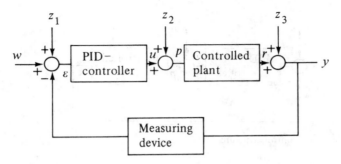

Fig. 8.1. Single-loop conventional PID control system.

tools for the evaluation of conventional control systems. Of course, there are many others which we shall not discuss here, such as stability margins, rise-time of response, rate of decay, heights of overshoots and their number, etc. If an integral criterion is used, the adjustable parameters of a PID-controller are chosen so that the criterion attains its minimum feasible value under the condition of asymptotic stability. Such values of the parameters are called optimal. More generally, the performance criterion creates an ordering in the space of parameters: one set of parameters is better than another if it gives a better (or smaller) value of the criterion, provided that the asymptotic stability is maintained.

According to the classical concept of feedback, the feedback loop should not change the signal, that is, the output y should be fed directly, with a negative sign, to the summing point on the input of the system. However, in many practical cases the dynamical distortions introduced by the measuring device and delays connected with the passage along the feedback loop cannot be neglected. The block 'Measuring Device' in Fig. 8.1 accounts for all such effects. If they can be neglected, we obtain the classical structure of conventional control (Fig. 8.2). Let us note that in the more general structure of Fig. 8.1, the error signal $w(t) - y(t)$ is not necessarily equal to $\varepsilon(t)$ (provided $z_1 = 0$, this is so in Fig. 8.2) and so we have to decide which of the two signals is to be used to form the integrand in an integral performance criterion.

We shall concentrate our further discussion on the system depicted in Fig. 8.2, extending the results where appropriate to the general structure of Fig. 8.1. The input of the system consists of the reference input w which normally is constant in time or, at most, is subject to step changes, and of disturbances z_1 and z_2. In our case, the disturbance z_1 represents an aggregate of input and output disturbances of the system and z_2 is the input disturbance of the controlled plant. One can consider disturbances of different types. They may have the character of delta functions—they then represent jumps of the trajectory, they may be step functions and, last but not least, they may be arbitrary, highly irregular functions of time. Especially in this last case, disturbances are often approached by methods of the theory of stochastic processes.

We shall now construct the mathematical model of the control system in the time domain. The controlled plant is described by a set of linear vector equations with time delays

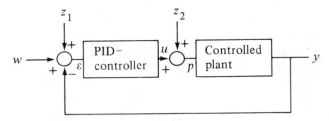

Fig. 8.2. Classical PID control structure.

$$\dot{x}(t) = \sum_{i=0}^{N} [A_i x(t - h_i) + B_i p(t - h_i)], \quad t \geq 0 \tag{8.2}$$

$$y(t) = \sum_{i=0}^{N} [C_i x(t - h_i) + D_i p(t - h_i)], \quad t \geq 0. \tag{8.3}$$

where $\dim x(t) = n$, $\dim y(t) = \dim p(t) = 1$, $0 = h_0 < h_1 < \ldots < h_N$. Time delays appear both in the differential equation (8.2) and in the output equation (8.3). To obtain a complete description we add the controller equation (8.1) and the equations of summing points

$$\left. \begin{array}{l} \varepsilon(t) = w(t) + z_1(t) - y(t) \\ p(t) = u(t) + z_2(t) \end{array} \right\} \text{ for } t \geq 0. \tag{8.4}$$
$$\tag{8.5}$$

The model (8.1)–(8.5) is linear, but rather complex. It includes an integro-differential equation (8.1), differential equations with time delays (8.2), a difference equation with time delays (8.3) and algebraic (static) equations (8.4) and (8.5). In order to calculate and analyse integral performance criteria we have either to find the Laplace transform of the error (see Chapters 3 and 4) or to reduce the model to one of the typical differential–difference systems discussed in Chapter 5. The latter approach makes it possible to use a well-established mathematical theory to study such fundamental qualitative questions as the existence and uniqueness of solutions and the problem of asymptotic stability, as well as to use existing formulae for solutions. This is important since such a theory does not exist for the original model (8.1)–(8.5). We shall thus aim at the reduction of system (8.1)–(8.5) to an equivalent, possibly simplest, typical mathematical problem.

To remove the integral in equation (8.1) we introduce a new variable ε_1,

$$\dot{\varepsilon}_1(t) = \varepsilon(t) \quad \text{for } t \geq 0, \quad \varepsilon_1(0) = \eta. \tag{8.6}$$

Let for $t < 0$

$$p(t) = a_R \varepsilon(t) + b_r \varepsilon_1(t) + c_R \dot{\varepsilon}(t) + z_2(t). \tag{8.7}$$

This is satisfied if, for example,

$$\eta = 0, \quad p(t) = z_2(t) \quad \text{and} \quad \varepsilon(t) = 0 \quad \text{for } t < 0. \tag{8.8}$$

Notice that (8.7) holds for all positive ts. Let us define new aggregate inputs

Sec. 8.1] **Time-delay systems with PID-controllers** 165

$$W_1(t) = \sum_{i=0}^{N} B_i z_2(t - h_i) \tag{8.9}$$

$$W_2(t) = w(t) + z_1(t) - \sum_{i=0}^{N} D_i z_2(t - h_i). \tag{8.10}$$

A simple reduction of variables and the substitution of (8.7) leads to the following system (for $t \geq 0$):

$$\dot{x}(t) = \sum_{i=0}^{N} [A_i x(t - h_i) + B_i a_R \varepsilon(t - h_i) + B_i b_R \varepsilon_1(t - h_i) + B_i c_R \dot{\varepsilon}(t - h_i)] + W_1(t) \tag{8.11}$$

$$\dot{\varepsilon}_1(t) = \varepsilon(t) \tag{8.12}$$

$$\varepsilon(t) = -\sum_{i=0}^{N} [C_i x(t - h_i) + D_i a_R \varepsilon(t - h_i) + D_i b_R \varepsilon_1(t - h_i) + D_i c_R \dot{\varepsilon}(t - h_i)] + W_2(t). \tag{8.13}$$

This set of $n + 2$ differential–difference equations is equivalent to the original system (8.1)–(8.5) on the subspace of its initial conditions determined by (8.6) and (8.7), and for special free terms (8.9) and (8.10). The mathematical type of system (8.11)–(8.13) depends on the values of its coefficients.

With the use of purely algebraic manipulation, this system can be reduced in some cases to one of the types discussed in Chapter 5, that is, retarded or neutral one. We shall consider three such cases. First, let us assume that the controller includes the differential action component ($c_R \neq 0$) and the input of the controlled plant affects its output also in a direct, static way ($D_0 \neq 0$), that is,

$$c_R D_0 \neq 0. \tag{8.14}$$

Equation (8.13) can then be solved with respect to $\dot{\varepsilon}(t)$ and the whole system is easily transformed into a neutral system of dimension $n + 2$. Notice that in this case we have to fix the value of $\varepsilon(0)$ to obtain a unique solution. Moreover, a unique solution exists for any choice of $\varepsilon(0)$. It can be proved that $\varepsilon(0)$ is uniquely determined by the original set of equations (8.1)–(8.5) and other initial conditions if and only if (8.14) does not hold, and $1 + a_R D_0 \neq 0$.

In the second case we assume

$$c_R = 0 \quad \text{and} \quad a_R D_0 \neq -1, \quad a_R D_i = 0 \quad \text{for } i = 1, \ldots, N, \tag{8.15a}$$

that is, the controller does not include the D-component. Equation (8.13) can then be solved with respect to $\varepsilon(t)$. The substitution of $\varepsilon(t)$ into (8.11) and (8.12) yields a differential–difference system of equations of retarded type, with a finite number of time delays.

In the third case,

$$c_R D_0 = 0 \quad \text{and} \quad B_i = 0, \quad D_i = C_i = 0 \quad \text{for } i = 1, \ldots, N. \tag{8.15b}$$

We introduce a new variable

$$z(t) = x(t) - B_0 c_R \varepsilon(t)$$

and substitute this into equations (8.11) and (8.13). If we additionally assume

$$1 + a_R D_0 + c_R C_0 B_0 \neq 0,$$

equation (8.13) can be solved with respect to $\varepsilon(t)$. Substituting this into (8.11) and (8.12) we again arrive at a retarded system of $n + 1$ differential–difference equations.

Example 8.1
The control system shown in Fig. 8.3 is described by the following equations (for $t \leq 0$):

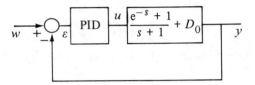

Fig. 8.3. System for Example 8.1.

$$u(t) = a_R \varepsilon(t) + b_R \varepsilon_1(t) + c_R \dot{\varepsilon}(t)$$

$$\dot{\varepsilon}_1(t) = \varepsilon(t)$$

$$\dot{x}(t) = -x(t) + u(t) + u(t-1)$$

$$y(t) = x(t) + D_0 u(t)$$

$$\varepsilon(t) = w(t) - y(t).$$

After the elimination of $u(t)$ and $y(t)$ we obtain a system corresponding to (8.11)–(8.13):

$$\dot{x}(t) = -x(t) + a_R \varepsilon(t) + b_R \varepsilon_1(t) + c_R \dot{\varepsilon}(t) + a_R \varepsilon(t-1)$$
$$+ b_R \varepsilon_1(t-1) + c_R \dot{\varepsilon}(t-1) \tag{8.16}$$

$$\dot{\varepsilon}_1(t) = \varepsilon(t) \tag{8.17}$$

$$\varepsilon(t) = -x(t) - D_0[a_R \varepsilon(t) + b_R \varepsilon_1(t) + c_R \dot{\varepsilon}(t)] + w(t). \tag{8.18}$$

By virtue of (8.7), this is valid for almost all $t \geq 0$. If $c_R D_0 \neq 0$, we calculate from the last equation

$$\dot{\varepsilon}(t) = \frac{1}{c_R D_0} [w(t) - x(t) - (a_R + c_R) D_0 \varepsilon(t) - b_R D_0 \varepsilon_1(t)]. \tag{8.19}$$

Substituting this into the first equation, we obtain

$$\dot{x}(t) = -\left(1 + \frac{1}{D_0}\right) x(t) - \frac{x(t-1)}{D_0} - c_R \varepsilon(t) - c_R \varepsilon(t-1) + \frac{w(t)}{D_0} + \frac{w(t-1)}{D_0}. \tag{8.20}$$

In the above derivation it has been additionally assumed that equation (8.4) holds also for $t < 0$, that is, $\varepsilon(t) = w(t) - y(t)$ for $t < 0$. We arrive at the conclusion that in the case $c_R D_0 \neq 0$ the system is described by a set of differential–difference equations of retarded type (8.17), (8.19), (8.20). Recall that in the general case assumption (8.14) allows us only to reduce system (8.16)–(8.18) to a neutral set of equations. This difference is due to the fact that the right-hand side of the output equation in our example, $y(t) = x(t) + d_0 u(t)$, does not contain delayed control terms (like $u(t-1)$).

Let us now assume $c_R = 0$ and $a_R D_0 \neq -1$. We then calculate from (8.18)

$$\varepsilon(t) = \frac{1}{a_R D_0 + 1} [w(t) - x(t) - b_R D_0 \varepsilon_1(t)] \tag{8.21}$$

and substitute this in (8.16) and (8.17)

$$\dot{x}(t) = \frac{1}{a_R D_0 + 1} [-(a_R D_0 + 1 + a_R) x(t) - a_R x(t-1) + b_R \varepsilon_1(t)$$
$$+ b_R \varepsilon_1(t-1) + a_R w(t) + a_R w(t-1)] \tag{8.22}$$

$$\dot{\varepsilon}_1(t) = \frac{1}{a_R D_0 + 1} [w(t) - x(t) - b_R D_0 \varepsilon_1(t)]. \tag{8.23}$$

We have again assumed $\varepsilon(t) = w(t) - y(t)$ for $t < 0$. Equations (8.22) and (8.23) are a differential–difference system of retarded type. Its solution gives the error signal by virtue of (8.21).

The reduction to a neutral or retarded-type system may be also achieved in some cases different from (8.14), (8.15a) or (8.15b), if a more ingenious transformation is used. This may be observed in our example if $c_R \neq 0$ and $D_0 = 0$. Let us assume that $\varepsilon(t) = -x(t) + w(t)$ also for negative ts and that this equality can be differentiated, that is,

$$\dot{\varepsilon}(t) = -\dot{x}(t) + \dot{w}(t) \quad \text{for every } t. \tag{8.24}$$

If $c_R \neq -1$, then system (8.16), (8.17), (8.24) can be easily solved with respect to the vector of derivatives $(\dot{x}(t), \dot{\varepsilon}(t), \dot{\varepsilon}_1(t))$, which yields

$$\dot{x}(t) + \frac{c_R}{c_R + 1} \dot{x}(t-1) = \frac{1}{c_R + 1} [-(1 + a_R) x(t) - a_R x(t-1) + b_R \varepsilon_1(t)$$
$$+ b_R \varepsilon_1(t-1) + a_R w(t) + a_R w(t-1) + c_R \dot{w}(t) + c_R \dot{w}(t-1)]$$

$$\dot{\varepsilon}_1(t) = -x(t) + w(t).$$

This is a neutral system of $n + 1$ differential–difference equations. It should be stressed that the reduction was obtained with the use of a non-algebraic transformation and was possible only under the additional assumption of differentiability of the function w and initial conditions.

We shall now calculate the Laplace transform $\hat{\varepsilon}(s)$ of the error signal ε. In this chapter it is more convenient for symbols with circumflexes to denote Laplace transforms, e.g. $\hat{\varepsilon}(s) = (\mathscr{L}\varepsilon)(s)$. Define the transfer function of the PID-controller

$$G_R(s) = a_R + \frac{b_R}{s} + c_R s. \tag{8.25}$$

The relationship between the output and the input of the controller is

$$\hat{u}(s) = G_R(s)\hat{\varepsilon}(s) + \frac{b_R}{s}\varepsilon_1(0) - c_R\varepsilon(0). \tag{8.26}$$

Of course $\varepsilon_1(0) = \eta$. We further define

$$A'(s) = \sum_{i=0}^{N} e^{-sh_i} A_i \tag{8.27}$$

and similarly $B'(s)$, $C'(s)$ and $D'(s)$. Let also

$$A(s,t) = e^{-st} \sum_{i=j}^{N} e^{-sh_i} A_i \quad \text{for } t \in [-h_j, -h_{j-1}), \ j = 1, \ldots, N. \tag{8.28}$$

We determine $B(s,t)$, $C(s,t)$ and $D(s,t)$ in an analogous way. Moreover, let

$$R(s) = [sI - A'(s)]^{-1}. \tag{8.29}$$

The Laplace transformation of equations (8.2)–(8.5) gives (arguments s and t are omitted for the sake of simplicity)

$$\hat{x} = R[B'\hat{p} + x(0) + \int_{-h_N}^{0} (Ax + Bp)\,dt]$$

$$\hat{y} = C'\hat{x} + D'\hat{p} + \int_{-h_N}^{0} (Cx + Dp)\,dt$$

$$\hat{\varepsilon} = \hat{w} + \hat{z}_1 - \hat{y}$$

$$\hat{p} = \hat{u} + \hat{z}_2.$$

Let us introduce the transfer function of the controlled plant

$$G = C'RB' + D',$$

the transfer function between the input of the control system and the input of the controller (we assume $G(s)G_R(s) \neq -1$ for some s)

$$G_1 = (1 + GG_R)^{-1}$$

and the transfer function between the input of the disturbance z_2 and the input of the controller

$$G_2 = -G_1 G.$$

Let also

$$H(s) = -G_1(s)C'(s)R(s)$$

$$H_1(s,t) = -G_1(s)[C'(s)R(s)A(s,t) + C(s,t)]$$

$$H_2(s,t) = -G_1(s)[C'(s)R(s)B(s,t) + D(s,t)].$$

After some algebraic manipulation we obtain the desired formula

$$\hat{\varepsilon} = G_1(\hat{w} + \hat{z}_1) + G_2\hat{z}_2 - G_2 C_R \varepsilon(0) + G_2 \frac{b_R}{s} \varepsilon_1(0) + Hx(0)$$

$$+ \int_{-h_N}^{0} (H_1 x + H_2 p)\, dt. \tag{8.30}$$

The expression for $\hat{\varepsilon}$ has been obtained in a formal fashion and so the meaning of the pointwise initial conditions $\varepsilon(0)$, $\varepsilon_1(0)$ and $x(0)$ needs some explanation. The quantities $\varepsilon(0)$ and $\varepsilon_1(0)$ are free parameters which are interpreted as left-hand limits at zero, $\varepsilon(0-)$ and $\varepsilon_1(0-)$, that is, they represent the state of the controller just before it begins to work. To give an interpretation for $x(0)$, denote by $x(t; \xi, p)$ the solution of equation (8.2) corresponding to initial condition ξ and input p. The mapping $(\xi, p) \mapsto x(t; \xi, p)$ is linear; its explicit form is given by the variation-of-constants formula. We substitute $x(0; \xi, 0)$ for the symbol $x(0)$ in (8.30), where ξ is the appropriate initial condition. If we confine our considerations to initial conditions such that the function $t \mapsto x(t; \xi, 0)$ is continuous at $t = 0$, then we may equivalently put $x(0-)$ for $x(0)$ in (8.30). It is worth recalling that there are theoretical and practical reasons for considering solutions of time-delay equations, discontinuous at $t = 0$. For example, the fundamental solution Φ of (8.2) satisfies $\Phi(0) = I$ and $\Phi(t) = 0$ for $t < 0$.

Let us note that using the Laplace transform technique we have obtained a universal and relatively simple expression for $\hat{\varepsilon}(s)$, contrary to the analysis in the time-domain where many particular cases must be separately considered. On the other hand, the study of pole location in (8.30), which is necessary for the stability analysis, is simpler if we know the type of the differential–difference system.

Example 8.2

For the system considered in Example 8.1 (Fig. 8.3) we have

$$G_R(s) = a_R + \frac{b_R}{s} + c_R s$$

$$G(s) = \frac{1 + e^{-s}}{1 + s} + D_0$$

$$G_1(s) = \frac{s(s+1)}{M_1(s) + M_2(s)e^{-s}},$$

where

$$M_1(s) = c_R D_0 s^3 + (c_R + c_R D_0 + a_R D_0 + 1)s^2$$
$$+ (b_R D_0 + a_R + a_R D_0 + 1)s + b_R(1 + D_0)$$

$$M_2(s) = c_R s^2 + a_R s + b_R,$$

$$G_2(s) = -G_1(s)G(s)$$

$$H(s) = -\frac{G_1(s)}{s+1}$$

$$H_2(s,t) = H(s)e^{-s(t+1)}.$$

Formula (8.30) takes the form

$$\hat{\varepsilon} = G_1 \hat{w} - G_2 c_R \varepsilon(0) + G_2 \frac{b_R}{s} \varepsilon_1(0) + Hx(0) + \int_{-1}^{0} H_2 u \, dt.$$

Let us return to the control system with non-direct feedback presented in Fig. 8.1 and assume that the dynamics of the measuring device and of the feedback loop is described by a system of time-delay equations, analogous to that which describes the controlled plant

$$\dot{\xi}(t) = \sum_{i=0}^{N} [A_{F_i} \xi(t - h_i) + B_{F_i} y(t - h_i)] \qquad (8.31)$$

$$v(t) = \sum_{i=0}^{N} [C_{F_i} \xi(t - h_i) + D_{F_i} y(t - h_i)]. \qquad (8.32)$$

To obtain a complete model we replace $y(t)$ by $r(t)$ in equation (8.3), add equations (8.31), (8.32) to (8.1), (8.2) and (8.3) and replace the equations of summing points (8.4) and (8.5) by the following:

$$\varepsilon(t) = w(t) + z_1(t) - v(t) \qquad (8.33)$$

$$p(t) = u(t) + z_2(t) \qquad (8.34)$$

$$y(t) = r(t) + z_3(t). \qquad (8.35)$$

A detailed analysis of this system of equations leads to results similar to those above. If $c_R D_0 D_{F_0} \neq 0$, then the system can be transformed into a neutral system of differential–difference equations. In this case we have to fix $\varepsilon(0)$ to have a unique solution. If $c_R D_0 D_{F_0} = 0$, we have to deal with many particular cases of which the following is most important:

$$c_R = 0, \quad a_R D_0 D_{F_0} \neq -1, \quad a_R D_i D_{F_j} = 0 \; \forall i,j : i+j > 0. \qquad (8.36)$$

In this case our system can be easily reduced to a differential–difference system of retarded type.

The derivation of Laplace transforms $\hat{\varepsilon}$ and $\hat{y}$ may be easily done in a similar manner as for the system with direct feedback. Define

$$A'_F(s) = \sum_{i=0}^{N} e^{-sh_i} A_{F_i}$$

$$A_F(s,t) = e^{-st} \sum_{i=j}^{N} e^{-sh_i} A_{F_i} \quad \text{for } t \in [-h_j, -h_{j-1}), \; j = 1, \ldots, N$$

and similarly B'_F, C'_F, D'_F and B_F, C_F, D_F. Let further

$$R_F(s) = [sI - A'_F(s)]^{-1}.$$

Define the transfer functions: between the input and the output of the measuring device

$$G_F(s) = C'_F(s)R_F(s)B'_F(s) + D'_F(s),$$

between the input of disturbances z_1 and the input of the controller (it is assumed that $G_F(s)G(s)G_R(s) \neq -1$ for some s)

$$G_1(s) = [1 + G_F(s)G(s)G_R(s)]^{-1},$$

between the input of disturbances z_2 and the input of the controller

$$G_2(s) = -G_1(s)G_F(s)G(s),$$

and between the input of disturbances z_3 and the input of the controller

$$G_3(s) = -G_1(s)G_F(s).$$

We also define the functions

$$H(s) = -G_1(s)G_F(s)C'(s)R(s)$$

$$H_F(s) = -G_1(s)C'_F(s)R_F(s)$$

$$H_1(s, t) = G_3(s)[C'(s)R(s)A(s, t) + C(s, t)]$$

$$H_2(s, t) = G_3(s)[C'(s)R(s)B(s, t) + D(s, t)]$$

$$H_{F_1}(s, t) = -G_1(s)[C'_F(s)R_F(s)A_F(s, t) + C_F(s, t)]$$

$$H_{F_2}(s, t) = -G_1(s)[C'_F(s)R_F(s)B_F(s, t) + D_F(s, t)].$$

In this notation the formula for the Laplace transform of ε takes the form

$$\hat{\varepsilon} = G_1(\hat{w} + \hat{z}_1) + G_2\hat{z}_2 + G_3\hat{z}_3 - G_2 c_R \varepsilon(0) + G_2 \frac{b_R}{s} \varepsilon_1(0) + Hx(0)$$

$$+ \int_{-h_N}^{0} (H_1 x + H_2 p) \, dt + H_F \xi(0) + \int_{-h_N}^{0} (H_{F_1}\xi + H_{F_2}y) \, dt. \tag{8.37}$$

The pointwise initial conditions $\varepsilon(0)$, $\varepsilon_1(0)$, $x(0)$ are understood as in (8.30) and the meaning of $\xi(0)$ is analogous as that of $x(0)$.

The formula for $\hat{y}$ is formally identical with (8.37) if we redefine the notations:

$$G_3 = (1 + GG_R G_F)^{-1}$$

$$G_2 = G_3 G$$

$$G_1 = G_3 G G_R$$

$$H = G_3 C' R.$$

8.2 OPTIMIZATION OF CONVENTIONAL CONTROL SYSTEMS WITH FIRST-ORDER PLANT AND UNIT STEP INPUT

We consider the simple conventional control system shown in Fig. 8.2 where the controller is described by equation (8.1) or the transfer function $G_R(s)$ (8.25) and the controlled plant is a first-order system with delay in its output, described by the following equations:

$$\left.\begin{array}{l}\dot{x}(t) = -qx(t) + Kp(t) \\ y(t) = x(t-h)\end{array}\right\} t \geq 0, \qquad (8.38)(8.39)$$

where $h > 0$, $q \geq 0$. The transfer function of the controlled plant is

$$G(s) = \frac{Ke^{-sh}}{s+q}. \qquad (8.40)$$

In order to express the Laplace transform of the error signal $\hat{\varepsilon}(s)$ we denote

$$\alpha = Ka_R, \quad \beta = Kb_R, \quad \gamma = Kc_R \qquad (8.41)$$

$$M(s) = s(s+q)e^{sh} + \gamma s^2 + \alpha s + \beta \qquad (8.42)$$

and

$$L(s) = s(s+q)e^{sh}(\hat{w}(s) + \hat{z}_1(s)) - sK\hat{z}_2(s) - sx(0)$$

$$-\beta\varepsilon_1(0) + s\gamma\varepsilon(0) - s(s+q)\int_{-h}^{0} e^{-st}x(t)\,dt. \qquad (8.43)$$

We then have

$$\hat{\varepsilon}(s) = \frac{L(s)}{M(s)}. \qquad (8.44)$$

As we have already mentioned, asymptotic stability of the closed-loop system is a basic requirement when searching for optimal controller settings. This requirement constitutes a constraint which determines the set of admissible values of a_R and/or b_R and/or c_R, depending on the regulator type. The conditions of asymptotic stability for our system can be obtained in an explicit form (see Chapter 2, and Gorecki, Fuksa, Grabowski and Korytowski (1989)). Let us first assume that the controller is of P or PD type so that $\beta = 0$. The system is asymptotically stable if and only if

(a) $\quad \alpha - q \leq 0 \quad$ and $\quad \alpha + q > 0, \quad -1 < \gamma < 1, \qquad (8.45)$

or

(b) $\quad \alpha - q > 0 \quad$ and $\quad \alpha + q > 0, \quad -1 < \gamma < 1, \quad$ and

$$h\sqrt{\frac{\alpha^2 - q^2}{1 - \gamma^2}} < \arccos\left(-\frac{q + \alpha\gamma}{q\gamma + \alpha}\right). \qquad (8.46)$$

If the controller is of PI or PID type, these conditions are slightly more complicated. The closed-loop system is asymptotically stable if and only if

$$\alpha + q > 0, \quad \beta > 0, \quad -1 < \gamma < 1 \tag{8.47}$$

and

$$h\sigma < \arccos \frac{\beta - q\alpha - \gamma\sigma^2}{q^2 + \sigma^2} \tag{8.48}$$

where

$$\left. \begin{aligned} \sigma &= \sqrt{\frac{\Delta - q^2 + \alpha^2 - 2\beta\gamma}{2(1 - \gamma^2)}} \\ \Delta &= \sqrt{(\alpha^2 - q^2)(\alpha^2 - q^2 - 4\beta\gamma) + 4\beta^2} \end{aligned} \right\} . \tag{8.49}$$

When we want to draw the boundary of the stability region in the space of parameters, a parametric formulation is often more useful than formulae (8.46) or (8.48). To this end we substitute $s = i\omega$, $\omega > 0$ in the characteristic quasipolynomial $M(s)$ (8.42) and equate it to zero. We thus obtain

$$\left. \begin{aligned} \beta &= \omega(q \sin \omega h + \omega \cos \omega h + \gamma\omega) \\ \alpha &= \omega \sin \omega h - q \cos \omega h \end{aligned} \right\} \omega > 0. \tag{8.50}$$

These formulae are valid for all types of controller; we only have to put appropriate parameters equal to zero. For example, in the case of a PD-controller we put $\beta = 0$ and obtain

$$\left. \begin{aligned} q \sin \omega h + \omega \cos \omega h + \gamma\omega &= 0 \\ \omega \sin \omega h - q \cos \omega h &= \alpha \end{aligned} \right\} \omega > 0. \tag{8.51}$$

The range of ω values in (8.50) or (8.51) is bounded from above by the value of ω at which the curve enters the region, inadmissible with respect to one of remaining constraints.

In the sections which follow we shall deal with the integral square error

$$J = \int_0^\infty \varepsilon(t)^2 \, dt \tag{8.52}$$

for the closed-loop control system of Fig. 8.2 with the controlled plant described by (8.38), (8.39) and (8.40). We shall consider the system activated by three different kinds of stimuli, namely unit step input, initial conditions, and white Gaussian noise. In each case we first give analytic formulae for J. We then use these to discuss the problem of optimal controller settings, which is stated as follows: find admissible values of decision variables a_R, b_R and c_R (or, equivalently, of α, β and γ) such that the corresponding value of J is minimal. All other factors which affect J, that is, inputs and initial conditions, are treated as fixed. The set of admissible decision variables is determined by the conditions of asymptotic stability and by the controller type. For example, if we confine ourselves to P-controllers, we then assume $b_R = c_R = 0$, or $\beta = \gamma = 0$.

We begin with the analysis of a control system driven by a unit step input. It is thus assumed that for all negative moments of time our system is in equilibrium with $x(t) = 0$, $y(t) = 0$, $w(t) = 0$, $\varepsilon(t) = 0$, $\varepsilon_1(t) = 0$, $u(t) = 0$, $z_1(t) = 0$ and $z_2(t) = 0$. At $t = 0$ a unit step function, $w(t) = H(t)$, is fed to the input and all later processes in the control system are reactions to this stimulus. In order to calculate the integral performance crtierion J (8.52) we use Parseval's theorem method of Chapter 4. To this end the Laplace transform of the error signal is needed. If we substitute zero initial conditions, zero disturbances and $\hat{w}(s) = 1/s$ in (8.43), (8.44), we obtain

$$\hat{\varepsilon}(s) = \frac{s+q}{s(s+q) + (\gamma s^2 + \alpha s + \beta)e^{-sh}} \tag{8.53}$$

for the system with PID-controller. In the case of PD-controller, $\hat{\varepsilon}(s)$ has a pole at zero unless $q = 0$ and $\alpha \ne 0$, and the integral in (8.52) does not exist, the steady-state error not being zero. We thus assume

$$q = 0 \quad \text{and} \quad \alpha \ne 0 \tag{8.54}$$

for the system with a PD-controller and obtain

$$\hat{\varepsilon}(s) = \frac{1}{s + (\gamma s + \alpha)e^{-sh}}. \tag{8.55}$$

The expressions for the integral performance criterion J are derived along the lines of Chapter 4. In the case of a system with PD-controller we have

$$J = \frac{1}{2(1-\gamma^2)\sigma} \frac{\sqrt{1-\gamma^2} + \sin \sigma h}{\gamma + \cos \sigma h} \tag{8.56}$$

where

$$\sigma = \frac{\alpha}{\sqrt{1-\gamma^2}}, \quad \alpha > 0. \tag{8.57}$$

For the system with a PID-controller we obtain

$$J = \frac{1}{2\Delta}\left[(q^2 - \rho^2)\frac{\gamma\rho^2 - \alpha q + \beta + (q^2 - \rho^2)\cosh \rho h}{(\alpha - q\gamma)\rho^2 - q\beta - \rho(q^2 - \rho^2)\sinh \rho h} \right.$$

$$\left. - (q^2 + \sigma^2)\frac{\gamma\sigma^2 + \alpha q - \beta - (q^2 + \sigma^2)\cos \sigma h}{(\alpha - q\gamma)\sigma^2 + q\beta - \sigma(q^2 + \sigma^2)\sin \sigma h}\right] \tag{8.58}$$

where σ and Δ are determined by (8.49) and

$$\rho = \sqrt{\frac{\Delta + q^2 - \alpha^2 + 2\beta\gamma}{2(1-\gamma^2)}}. \tag{8.59}$$

Formula (8.58) is valid for $\beta \neq q(\alpha - \gamma q)$. If $\beta = q(\alpha - \gamma q)$, the first term in the square brackets is an indeterminate of the type $0/0$ and the value of J is to be calculated either by a limit procedure ($\lim J$ as $\beta \to q(\alpha - \gamma q)$), or separately by the direct approach. The calculations give

$$J = \frac{1}{2(1-\gamma^2)\sigma} \frac{\sqrt{1-\gamma^2} + \sin\sigma h}{\gamma + \cos\sigma h}, \quad \sigma = \frac{\alpha - \gamma q}{\sqrt{1-\gamma^2}} \tag{8.60}$$

for $\beta = q(\alpha - \gamma q)$.

We shall now discuss optimal settings for different types of controllers.

P-controller

The closed-loop system is asymptotically stable if and only if $\alpha h \in (0, \pi/2)$. We use formula (8.56) with $\gamma = 0$. From the condition

$$\frac{dJ}{d\alpha} = 0$$

we easily find that α_{opt}, the optimal value of α, satisfies the equation

$$\alpha_{opt} h = \cos \alpha_{opt} h. \tag{8.61}$$

Hence $\alpha_{opt} h \approx 0.7391$, which gives $J_{opt} \approx 1.532 h$. Notice that J tends to infinity as αh approaches the boundaries of the stability interval $(0, \pi/2)$.

PD-controller

Figure 8.4 shows the dependence of the performance index J on α and γ in the whole of the stability region, for the delay h fixed at 1. J tends to infinity as (α, γ) approaches the boundary of the stability region and is a well-behaved, smooth function in the interior of this region. It has there a unique minimum. In order to discuss optimal settings of a PD-controller, it is convenient to rewrite formula (8.56) as follows:

$$J = \frac{h}{2\Theta \sin^2 \rho} \tan \frac{\rho + \Theta}{2} \tag{8.62}$$

where

$$\Theta = \sigma h = \frac{\alpha h}{\sqrt{1-\gamma^2}}, \quad \rho = \arccos \gamma. \tag{8.63}$$

We minimize J subject to constraints resulting from the stability conditions (8.46)

$$0 < \rho < \pi, \quad 0 < \Theta < \pi - \rho. \tag{8.64}$$

From the conditions

$$\frac{\partial J}{\partial \Theta} = 0, \quad \frac{\partial J}{\partial \rho} = 0$$

we easily find that the optimal values of Θ and ρ satisfy

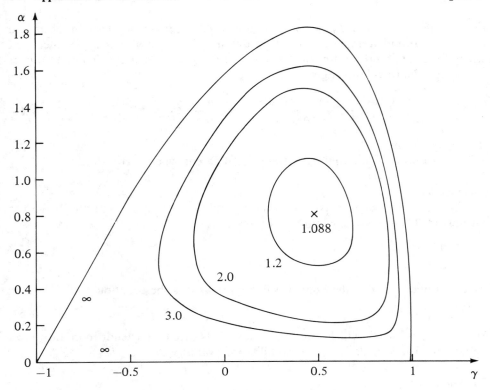

Fig. 8.4. Cost contours for the PD-controller and unit step input.

$$\left.\begin{array}{l}\Theta = \sin(\rho + \Theta) \\ \Theta = \tfrac{1}{2} \tan \rho. \end{array}\right\} \quad (8.65)$$

This is a transcendental set of equations and must be solved numerically. The elimination of ρ yields

$$\Theta\sqrt{1 + 4\Theta^2} = 2\Theta \cos \Theta + \sin \Theta. \quad (8.66)$$

Hence we obtain the optimal value of Θ,

$$\Theta_{\text{opt}} \cong 0.9151.$$

The second of equations (8.65) gives the optimal settings

$$\gamma_{\text{opt}} = \frac{1}{\sqrt{1 + 4\Theta_{\text{opt}}^2}} \approx 0.4795$$

$$\alpha_{\text{opt}} h = \frac{2\Theta_{\text{opt}}^2}{\sqrt{1 + 4\Theta_{\text{opt}}^2}} \approx 0.803.$$

The optimal value of the performance index is

$$J_{opt} \approx 1.088h.$$

PI-controller

To simplify further considerations, let us notice that we may assume $h = 1$ without any loss of generality. Indeed, if we denote the left-hand side of (8.58) by $J(\alpha, \beta, \gamma, q, h)$ we have the obvious identity (for $h > 0$)

$$J(\alpha, \beta, \gamma, q, h) = hJ(\alpha h, \beta h^2, \gamma, qh, 1). \tag{8.67}$$

From now on it will be thus assumed $h = 1$. The expression for the performance index is obtained from (8.58) after putting $\gamma = 0$, $h = 1$

$$J = \frac{1}{2\Delta}\left[(q^2 - \rho^2)\frac{\beta - \alpha q + (q^2 - \rho^2)\cosh\rho}{\alpha\rho^2 - q\beta - \rho(q^2 - \rho^2)\sinh\rho} \right.$$
$$\left. + (q^2 + \sigma^2)\frac{\beta - \alpha q + (q^2 + \sigma^2)\cos\sigma}{\alpha\sigma^2 + q\beta - \sigma(q^2 + \sigma^2)\sin\sigma}\right] \tag{8.68}$$

where

$$\Delta = \sqrt{(\alpha^2 - q^2)^2 + 4\beta^2}$$
$$\rho = \sqrt{\frac{\Delta + q^2 - \alpha^2}{2}}, \quad \sigma = \sqrt{\frac{\Delta - q^2 + \alpha^2}{2}}. \tag{8.69}$$

The asymptotic stability conditions are given by (8.47), (8.48):

$$\alpha + q > 0, \quad \beta > 0$$
$$\sigma < \arccos\frac{\sigma^2 - \alpha^2}{q\alpha + \beta}. \tag{8.70}$$

The dependence of J on α and β for a system with a PI-controller and an astatic plant (with $q = 0$) is depicted in Fig. 8.5. It is worth noting that J tends to infinity when (α, β) approaches that part of the boundary of the stability region which is given by

$$\alpha = \omega\sin\omega, \quad \beta = \omega^2\cos\omega, \quad \omega \in \left(0, \frac{\pi}{2}\right). \tag{8.71}$$

For every fixed admissible α, J decreases with decreasing β and so the problem of optimal settings for $q = 0$ has no solution in the region of asymptotic stability.

If q is positive, that is, the controlled plant is inertial, then J tends to infinity on the whole of the boundary of the region of asymptotic stability and has a unique minimum in the interior of this region. Fig. 8.6 shows the dependence of J on α and β for $q = 1$. In Fig. 8.7 the optimal settings α_{opt} and β_{opt}, as well as the optimal value of the performance criterion J_{opt}, are plotted against the parameter q. The optimal values were found by numerical techniques.

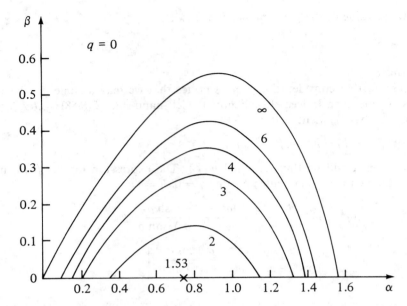

Fig. 8.5. Cost contours for PI-controller and astatic plant ($q = 0$), with unit step input.

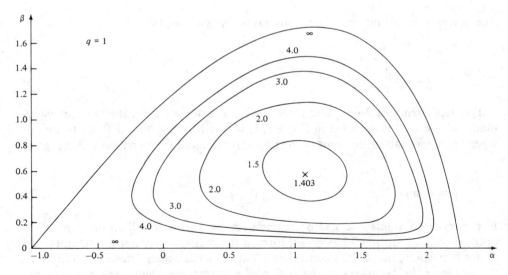

Fig. 8.6. Cost contours for inertial plant ($q = 1$) with PI-controller and unit step input.

PID-controller

For every $q > 0$ (inertial controlled plant) the performance criterion J (8.58) or (8.60) tends to infinity as (α, β, γ) approaches the boundary of the asymptotic stability region.

Sec. 8.2] Conventional control systems with step input 179

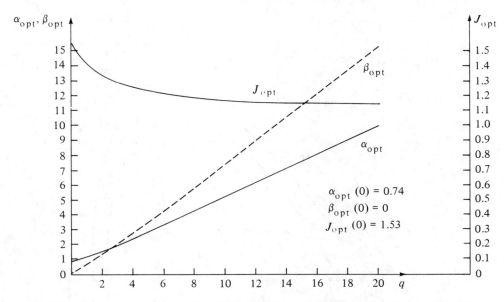

Fig. 8.7. Optimal values of α and β and the optimal cost, for PI-controller and unit step input.

Inside this region it is a well behaved, smooth function with a unique minimum which determines the optimal settings. If $q = 0$ (astatic plant), our optimization problem has no solution since J is a strictly increasing function of β and so does not attain its minimum for $\beta > 0$. Fig. 8.8 shows the dependence of J on α and β for fixed $q = 0$

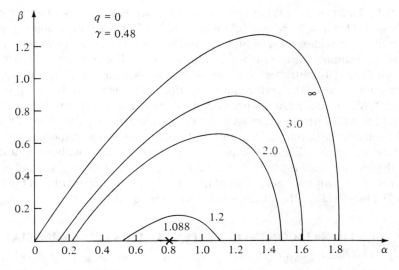

Fig. 8.8. Cost contours when $q = 0$ and $\gamma = 0.48$, for PID controller with unit step input.

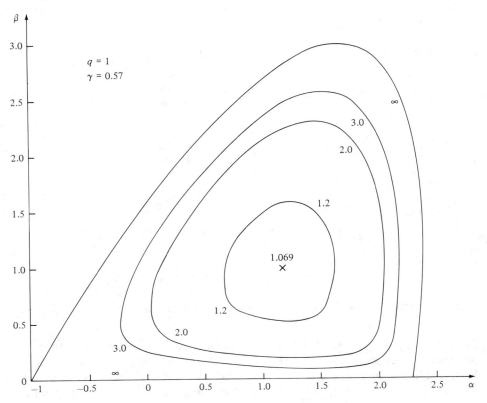

Fig. 8.9. Cost contours for $q = 1$ and $\gamma = 0.57$, for PID-controller with unit step input.

and $\gamma = 0.48$. This value of γ is the limit to which γ_{opt} tends as $q \to 0+$. The dependence of J on α and β for $q = 1$ and $\gamma = 0.57$ is depicted in Fig. 8.9. This value of γ is equal to γ_{opt} for $q = 1$. A comparison of Figs. 8.8 and 8.9 with Figs. 8.5 and 8.6, respectively, allows us to estimate how much the addition of a D-component improves the performance of a PI-controller. The plot of optimal settings, α_{opt}, β_{opt} and γ_{opt}, and of the optimal value of the performance functional J_{opt} as a function of q, is shown in Fig. 8.10. All optimal values where obtained by numerical methods.

Figs. 8.11 and 8.12 permit the most direct comparison of the performance of various types of controller with optimal settings. Both figures show the responses $y(t)$ to unit step inputs $w(t) = H(t)$. Fig. 8.11 concerns the control system with an astatic plant ($q = 0$); the responses are drawn for optimal P- and PD-controllers. Fig. 8.12 concerns the control system with an inertial plant ($q = 1$); the responses are drawn for optimal PI- and PID-controllers. In all cases the delay h is equal to 1.

8.3 CONVENTIONAL CONTROL SYSTEM ACTIVATED BY INITIAL CONDITIONS

We now consider the conventional control system of section 8.2 with non-zero initial

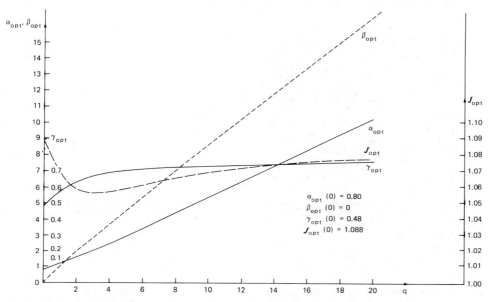

Fig. 8.10. Optimal values of α, β, γ and corresponding optimal cost for PID-controller and unit step input.

conditions and with the input w and disturbances z_1, z_2 equal to zero.

Let us first consider initial conditions of the form

$$x(0) = 1, \quad x(t) = 0 \quad \text{for } t < 0, \quad \varepsilon(0) = \varepsilon_1(0) = 0. \tag{8.72}$$

Notice that if $c_R \neq 0$, the system model (8.1), (8.38), (8.39), (8.4), (8.5) must be understood in a distributional sense, as ε is discontinuous at $t = kh$ (k positive integers), and in consequence u has Dirac's delta components at $t = kh$. In any case, however, ε is a well-defined, locally square integrable function. Formulae (8.43) and (8.44) give

$$\hat{\varepsilon}(s) = \frac{-s}{s(s+q)e^{sh} + \gamma s^2 + \alpha s + \beta}. \tag{8.73}$$

Similarly as in the previous section, also in this case we use Parseval's theorem method to calculate J (8.52). For $\beta \neq 0$ we obtain

$$J = -\frac{1}{2\Delta}\left[\rho^2 \frac{\gamma\rho^2 - \alpha q + \beta - (\rho^2 - q^2)\cosh \rho h}{(\alpha - q\gamma)\rho^2 - q\beta + \rho(\rho^2 - q^2)\sinh \rho h} \right.$$
$$\left. + \sigma^2 \frac{\gamma\sigma^2 + \alpha q - \beta - (\sigma^2 + q^2)\cos \sigma h}{(\alpha - q\gamma)\sigma^2 + q\beta - \sigma(\sigma^2 + q^2)\sin \sigma h}\right]. \tag{8.74}$$

In the case where $\beta = 0$, formula (8.73) reduces to

$$\hat{\varepsilon}(s) = \frac{-1}{(s+q)e^{sh} + \gamma s + \alpha} \tag{8.75}$$

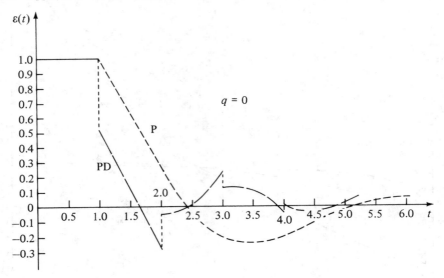

Fig. 8.11. Step responses for P and PD control at optimality for $q = 0$.

and we distinguish between three cases. Denote

$$\tau = \sqrt{\frac{|\alpha^2 - q^2|}{1 - \gamma^2}}. \tag{8.76}$$

If $\alpha^2 < q^2$, we have

$$J = \frac{1}{2(1 - \gamma^2)\tau} \frac{\gamma\tau^2 - q\alpha + (q^2 - \tau^2)\cosh \tau h}{(q\gamma - \alpha\tau + (q^2 - \tau^2)\sinh \tau h}. \tag{8.77}$$

If $\alpha^2 > q^2$,

$$J = \frac{1}{2(1 - \gamma^2)\tau} \frac{\gamma\tau^2 + q\alpha - (q^2 + \tau^2)\cos \tau h}{(q\gamma - \alpha)\tau + (q^2 + \tau^2)\sin \tau h}. \tag{8.78}$$

If $\alpha = q \neq 0$, we obtain

$$J = \frac{1 - \gamma + qh}{4(1 - \gamma^2)q}. \tag{8.79}$$

Let us notice that if $\beta = q(\alpha - q\gamma)$, the first fractional term in the square brackets in the right-hand side of (8.74) becomes indeterminate. To compute its value we divide the numerator and denominator by $\rho^2 - q^2$ and pass to the limit $\beta \to q(\alpha - q\gamma)$. This way we obtain for $\beta = q(\alpha - q\gamma)$

$$J = \frac{1}{2\Delta}\left(q^2 \frac{\cosh qh - \gamma}{\alpha - q\gamma + q \sinh qh} + \sigma \frac{\cos \sigma h - \gamma}{\sqrt{1 - \gamma^2} - \sin \sigma h}\right) \tag{8.80}$$

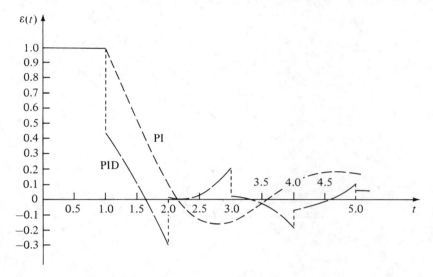

Fig. 8.12. Step responses for PI and PID control at optimality for $q = 1$.

where

$$\Delta = \alpha^2 + q^2 - 2\alpha\gamma q, \quad \sigma = \frac{\beta}{q\sqrt{1-\gamma^2}}. \tag{8.81}$$

The analysis of optimal settings implies that the addition of the I-component in the controller makes the performance functional J increase. We therefore confine our considerations to P- and PD-controllers.

P-controller
We substitute $\gamma = 0$ in formulae (8.76)–(8.79). Fig. 8.13 shows the dependence of J (8.52) on α and q in the whole of the stability region and in Fig. 8.14, the optimal values of α and J, α_{opt} and J_{opt} are drawn as functions of the parameter q. Both figures are drawn for the delay h fixed at 1. The results for other delays may be easily obtained using the identity $J(\alpha, q, h) = hJ(\alpha h, qh, 1) \, \forall \, h > 0$. Notice that in the case where $q = 0$, the optimal values of α and J are identical to those of section 8.2.

PD-controller
The dependence of the optimal settings α_{opt} and γ_{opt}, and of the optimal value of the functional J_{opt} on the parameter q, is depicted in Fig. 8.15. Again, if $q = 0$, the optimal values are identical to those of section 8.2. It is assumed that $h = 1$. The results for other delays are given by

$$J(\alpha, \gamma, q, h) = hJ(\alpha h, \gamma, qh, 1), \quad \forall \, h > 0.$$

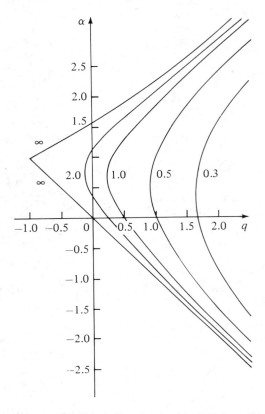

Fig. 8.13. Cost contours for P control, and point-wise initial conditions.

We shall now discuss the dependence of the integral functional J (8.52) on initial conditions. For the sake of simplicity we assume that the controller is of PD type. As all inputs are equal to zero, the system equations take the form (for $t \geq 0$)

$$u(t) = a_R \varepsilon(t) + c_R \dot{\varepsilon}(t) \tag{8.82}$$

$$\dot{x}(t) + qx(t) = Ku(t) \tag{8.83}$$

$$y(t) = -\varepsilon(t) = x(t-h). \tag{8.84}$$

Hence we immediately obtain a neutral equation for x,

$$\dot{x}(t) + \gamma \dot{x}(t-h) + qx(t) + \alpha x(t-h) = 0, \quad t \geq 0. \tag{8.85}$$

The performance functional is given by

$$J = \int_0^\infty \varepsilon(t)^2 \, dt = \int_{-h}^0 x(t)^2 \, dt + \int_0^\infty x(t)^2 \, dt. \tag{8.86}$$

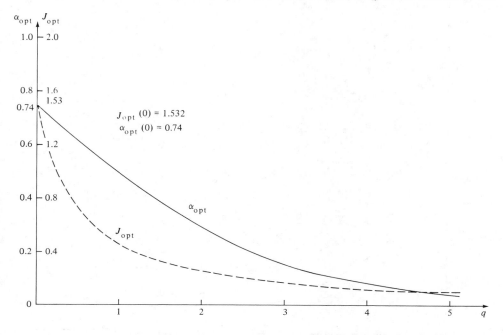

Fig. 8.14. Optimal values of α and cost for P control, and point-wise initial conditions.

We use the Lyapunov theory of Chapter 5 to calculate

$$J_1 = \int_0^\infty x(t)^2 \, dt. \tag{8.87}$$

Using formula (5.253) we find out that the performance functional J for arbitrary initial conditions is given by the following expression:

$$J = Py_0^2 + 2y_0 \int_{-h}^0 K(t+h)x(t)\,dt + \frac{1}{1-\gamma^2} \int_{-h}^0 x(t)^2\,dt$$

$$- 2 \int_{-h}^0 \int_t^0 [\gamma \dot{K}(\tau - t) + \alpha K(\tau - t)]x(t)x(\tau)\,d\tau\,dt, \tag{8.88}$$

where

$$y_0 = x(0) + \gamma x(-h). \tag{8.89}$$

It is assumed that x is right-continuous at $t = 0$ and $t = -h$. Provided explicit expressions for P and $K: [0, h] \to \mathbb{R}^1$ are known, formula (8.88) is a powerful tool

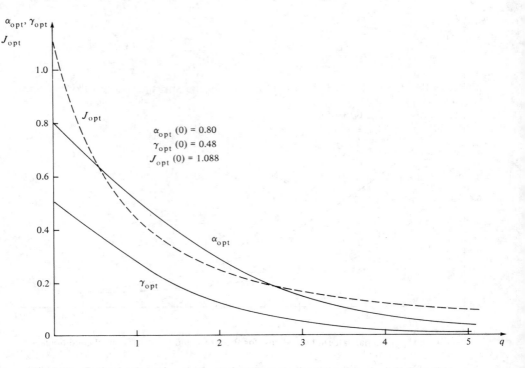

Fig. 8.15. Optimal values of α, γ and cost for PD-controller and point-wise initial conditions.

for the analysis of the performance functional, as it reduces the problem of computing values of J to simple quadratures. In order to calculate P and K we use equations (5.261)–(5.264). Notice that in our case $A_0 = 1$, $A_1 = \gamma$, $B_0 = q$, $B_1 = \alpha$, $V = 1$ and so this system takes the form

$$(1 - \gamma^2)Q = 1 \tag{8.90}$$

$$qP - K(h) = \tfrac{1}{2}Q \tag{8.91}$$

$$(\alpha + \gamma q)P + K(0) - \gamma K(h) = 0 \tag{8.92}$$

$$\dot{K}(t) + \gamma \dot{K}(h - t) + qK(t) + \alpha K(h - t) = 0, \quad t \in [-h, 0]. \tag{8.93}$$

The first three equations give

$$Q = \frac{1}{1 - \gamma^2} \tag{8.94}$$

$$qP - K(h) = \frac{1}{2(1 - \gamma^2)} \tag{8.95}$$

$$\alpha P + K(0) = \frac{-\gamma}{2(1 - \gamma^2)} \tag{8.96}$$

and

$$qK(0) + \alpha K(h) = -\frac{\alpha + q\gamma}{2(1 - \gamma^2)}. \tag{8.97}$$

We solve the differential equation (8.93) by using the approach described in Chapter 5. We introduce new variables

$$y_1(t) = \tfrac{1}{2}[K(t) + K(h - t)] \tag{8.98}$$

$$y_2(t) = \tfrac{1}{2}[K(t) - K(h - t)]. \tag{8.99}$$

By virtue of equation (8.93) they satisfy

$$\dot{y}_1 = \frac{\alpha - q}{1 - \gamma} y_2 \tag{8.100}$$

$$\dot{y}_2 = -\frac{\alpha + q}{1 + \gamma} y_1. \tag{8.101}$$

Denote

$$\sigma = \sqrt{\frac{|\alpha^2 - q^2|}{1 - \gamma^2}}. \tag{8.102}$$

We distinguish between three cases, which will be labelled A, B and C, and analysed in this order.

A. $\alpha^2 - q^2 < 0$

The general solution of (8.100), (8.101) is

$$y_1(t) = (\alpha - q)(k_1 e^{\sigma t} + k_2 e^{-\sigma t}) \tag{8.103}$$

$$y_2(t) = (1 - \gamma)\sigma(k_1 e^{\sigma t} - k_2 e^{-\sigma t}) \tag{8.104}$$

where k_1 and k_2 are arbitrary constants. Hence, by virtue of (8.98), (8.99)

$$K(t) = k_1[\alpha - q + (1 - \gamma)\sigma]e^{\sigma t} + k_2[\alpha - q - (1 - \gamma)\sigma]e^{-\sigma t} \tag{8.105}$$

$$K(h - t) = k_1[\alpha - q - (1 - \gamma)\sigma]e^{\sigma t} + k_2[\alpha - q + (1 - \gamma)\sigma]e^{-\sigma t}. \tag{8.106}$$

Since both these formulae represent the same function, we have

$$k_2 = k_1 e^{\sigma h}.$$

After introducing a new arbitrary constant K_1 we obtain the general solution of equation (8.93):

$$K(t) = K_1\left[(\alpha - q)\cosh \sigma\left(t - \frac{h}{2}\right) + (1 - \gamma)\sigma \sinh \sigma\left(t - \frac{h}{2}\right)\right]. \tag{8.107}$$

K_1 is calculated from the boundary condition (8.97):

$$K_1 = \frac{\alpha + q\gamma}{2(1 - \gamma^2)(q - \alpha)} \cdot \frac{1}{(\alpha + q)\cosh\sigma\frac{h}{2} + (1 - \gamma)\sigma \sinh\sigma\frac{h}{2}}. \qquad (8.108)$$

Thus

$$K(h) = \frac{\alpha + q\gamma}{2(1 - \gamma^2)(q - \alpha)} \cdot \frac{(\alpha - q)\cosh\sigma\frac{h}{2} + (1 - \gamma)\sigma \sinh\sigma\frac{h}{2}}{(\alpha + q)\cosh\sigma\frac{h}{2} + (1 - \gamma)\sigma \sinh\sigma\frac{h}{2}}. \qquad (8.109)$$

We further calculate from (8.95)

$$P = \frac{1}{2(q^2 - \alpha^2)} \cdot \frac{q^2 - \alpha^2 + (\alpha + \gamma q)\sigma \sinh\sigma h}{q + \alpha\gamma + (\alpha + \gamma q)\cosh\sigma h}. \qquad (8.110)$$

Notice that this is just another equivalent form of expression (8.77).

B. $\alpha^2 - q^2 > 0$

We could now repeat all the calculations with the characteristic roots equal to $\pm i\sigma$ but let us notice that it is sufficient to put $i\sigma$ in place of σ in the final formulae of case A. We thus obtain

$$K(t) = \frac{\alpha + q\gamma}{2(1 - \gamma^2)(q - \alpha)} \cdot \frac{(\alpha - q)\cos\sigma\left(t - \frac{h}{2}\right) - (1 - \gamma)\sigma \sin\sigma\left(t - \frac{h}{2}\right)}{(\alpha + q)\cos\sigma\frac{h}{2} - (1 - \gamma)\sigma \sin\sigma\frac{h}{2}} \qquad (8.111)$$

$$P = \frac{1}{2(\alpha^2 - q^2)} \cdot \frac{\alpha^2 - q^2 + (\alpha + \gamma q)q \sin\sigma h}{q + \alpha\gamma + (\alpha + \gamma q)\cos\sigma h}. \qquad (8.112)$$

This expression is equivalent to (8.78).

C. $\alpha^2 = q^2$

As the system is not asymptotically stable if $\alpha = -q$, we assume $\alpha = q \neq 0$. Equations (8.100) and (8.101) take the form

$$\dot{y}_1 = 0 \qquad (8.113)$$

$$\dot{y}_2 = -\frac{2q}{1 + \gamma} y_1. \qquad (8.114)$$

The general solution is

$$y_1 = (1 + \gamma)k_1 \qquad (8.115)$$

$$y_2 = -2k_1 q t + k_2 \qquad (8.116)$$

where k_1 and k_2 are arbitrary constants. The procedure used in case A therefore gives

$$K(t) = \frac{2qt - qh - \gamma - 1}{4(1 - \gamma^2)} \qquad (8.117)$$

and

$$P = \frac{1 - \gamma + qh}{4(1 - \gamma^2)q}. \qquad (8.118)$$

This result can also be directly obtained from the formulae of case A or B by an appropriate limit procedure.

Let us now compute the kernel in the double integral in formula (8.88).

A. $\alpha^2 - q^2 < 0$

In this case

$$\gamma \dot{K}(t) + \alpha K(t) = C\left[(\alpha - q) \cosh \sigma\left(t - \frac{h}{2}\right) + (1 + \gamma)\sigma \sinh \sigma\left(t - \frac{h}{2}\right)\right] \qquad (8.119)$$

where

$$C = \frac{\alpha - \gamma q}{1 + \gamma} K_1, \qquad (8.120)$$

K_1 given by (8.108).

B. $\alpha^2 - q^2 > 0$

In this case

$$\gamma \dot{K}(t) + \alpha K(t) = C\left[(\alpha - q) \cos \sigma\left(t - \frac{h}{2}\right) - (1 + \gamma)\sigma \sin \sigma\left(t - \frac{h}{2}\right)\right] \qquad (8.121)$$

where

$$C = \frac{\alpha^2 - \gamma^2 q^2}{2(1 + \gamma)(1 - \gamma^2)(q - \alpha)} \cdot \frac{1}{(\alpha + q)\cos \sigma \frac{h}{2} - (1 - \gamma)\sigma \sin \sigma \frac{h}{2}}. \qquad (8.122)$$

C. $\alpha = q \neq 0$

We then have

$$\gamma \dot{K}(t) + \alpha K(t) = \frac{q(2qt - qh + \gamma - 1)}{4(1 - \gamma^2)}. \qquad (8.123)$$

It is worth noting that in every case the kernel $\gamma \dot{K}(\tau - t) + \alpha K(\tau - t)$ can be represented as a sum of products of elementary functions of the variable τ and of elementary functions of the variable t, which considerably simplifies computing the double integral in (8.88).

8.4 APPLICATION TO STOCHASTIC INPUTS

In previous chapters we have considered various methods for the calculation of integral performance criteria of the form

$$J = \int_0^\infty e(t)\,dt \qquad (8.124)$$

where $e(t)$ is a squared error signal or squared response to initial conditions, or a combination of both. We have also considered modified signals, e.g. multiplied by simple functions of time, differentiated, etc. It is evident that such a criterion is useful only if $e(t)$ vanishes with time, $e(t) \to 0$ as $t \to \infty$. This confines possible application.

A slightly different integral performance criterion

$$J = \lim_{T \to \infty} \frac{1}{T} \int_0^T e(t)\,dt \qquad (8.125)$$

may be applied to the evaluation of control systems with irregularly varying disturbances of constant intensity, incessantly acting on the input. The output (or error) then does not tend to any limit but remains uniformly bounded and so the criterion (8.125) may have a finite, positive value. Of course, this form of criterion may also be used to evaluate how a system transmits (or dampens) a deterministic, e.g. sinusoidal, input but it proves most useful in analysis of stochastic disturbances. In particular, after an appropriate adaptation it gives simple formulae for the estimation of how stationary white Gaussian noise affects the output.

We shall begin with recalling some elements of the theory of stochastic systems and then we shall show how the results of earlier chapters may be used to calculate quadratic performance criteria for systems driven by stationary white Gaussian noise. Let us first consider a system without delays, described by a stationary linear differential equation

$$\dot{x}(t) = Ax(t) + Bu(t), \quad t \geq 0 \qquad (8.126)$$

where $x(t)$ is an n-dimensional state vector, $u(t)$ is an m-dimensional input vector, A and B are constant matrices of compatible dimensions. We shall deal with system (8.126) with stochastic inputs. Let a stochastic process $U: [0, \infty) \to R^m$, defined on the probabilistic space $(\Omega, \mathscr{F}, \mathscr{P})$, be an input of system (8.126). This means that for every fixed element ω of the space of elementary events Ω we treat the corresponding realization of the process $U(\cdot, \omega)$ as an input in equation (8.126), that is, $u(t) = U(t, \omega)$, $\forall t \geq 0$. By $\mathscr{F}$ we denote the space of stochastic events, that is, $\mathscr{F}$ is a σ-algebra of subsets of Ω, and $\mathscr{P}$ is the probability function, that is, a countably additive function from $\mathscr{F}$ into $[0, 1]$ such that $\mathscr{P}(\Omega) = 1$.

It is well known that for every integrable input u the corresponding solution of equation (8.126) is expresseed by means of the variation-of-constants formula (1.8):

$$x(t) = \Phi(t)x(0) + \int_0^t \Phi(t-s)Bu(s)\,ds \qquad (8.127)$$

where

$$\Phi(t) = \exp(At). \qquad (8.128)$$

Unfortunately, in most cases neither the differential equation (8.126) nor the solution formula (8.127) has a direct meaning when u is a stochastic noise. This results from the fact that almost all realizations of a stochastic process may be too irregular to be integrable.

A particularly important example of a stochastic process with this intriguing property is white Gaussian noise. This process is important both from the theoretical and practical points of view; for example, it gives a very good though idealized description of such real physical effects as thermal voltage (voltage resulting from thermal vibrations in the material of resistive elements). From the mathematical point of view, a stationary white Gaussian noise is a stochastic process whose value at every moment of time is a Gaussian stochastic variable, and the values at different moments of time are independent (therefore uncorrelated). The mean value and covariance matrix are constant. We assume that the former is equal to zero and the latter to Z, $Z = Z^T \geq 0$. The mathematical theory of system (8.126) driven by white Gaussian noise requires special constructions and a special understanding of equation (8.126). First, the so-called Wiener process, $W(t)$, is defined. This is a stochastic process with the following properties. For any sequence of time moments $0 \leq t_1 < t_2 < \ldots < t_k$ the corresponding increments $W(t_2) - W(t_1), W(t_3) - W(t_2), \ldots, W(t_k) - W(t_{k-1})$ are independent Gaussian stochastic variables with zero mean value and covariance proportional to the increment of time:

$$E([W(t) - W(s)][W(t) - W(s)]^T) = (t-s)Z, \quad \forall t, s, \quad t \geq s. \qquad (8.129)$$

It is assumed $W(0) = 0$. The Wiener process is a mathematical, idealized model of the Brownian motion, that is, the motion of a small particle of dust or pollen in a liquid, caused by thermal motion of molecules of the liquid. It is assumed that the particle is much heavier than the molecules of liquid and the molecules move in a purely chaotic way. The process of transferring momentum to and from the particle, i.e. its acceleration, may be treated as a stationary white Gaussian noise. It is intuitively clear that increments of momentum of the particle are independent Gaussian variables with zero mean value and variance proportional to the increment of time. If we assume that the particle was at rest at the beginning, then its speed can be approximated by a Wiener process. Let us notice that the relationship between a white Gaussian noise and the corresponding Wiener process is analogous to the relationship between an integrable function and its integral.

The stochastic integral is the next important notion we need. Let us first recall the definition of limit in the sense of quadratic mean. We say that the stochastic process $X:[0, \infty) \to \mathbb{R}^m$ has a limit Y at moment t_0 (Y is a stochastic variable), if the quadratic mean of the difference $X(t) - Y$ tends to zero as t tends to t_0:

$$E|X(t) - Y|^2 \to 0 \quad \text{as } t \to t_0.$$

$|\cdot|$ denotes the Euclidean norm. Once we have defined the limit, we are able to introduce for stochastic processes all concepts based on this definition, such as continuity, derivatives, integrals which are determined in an analogous way as in the classic calculus for vector functions of real variable. In particular, analogues of the Riemann, Stieltjes–Riemann, Lebesgue and Stieltjes–Lebesgue integrals are defined. Let f be a deterministic, continuous matrix-valued function from $[0, T]$ into $R^{m \times n}$ and let X be a stochastic process with its values in R^n. The definite Stieltjes–Riemann integral of f with respect to X, denoted by

$$Y = \int_0^T F(t) \, dX(t), \tag{8.130}$$

is defined as the limit of the sequence of partial sums, that is,

$$E | \sum_{i=0}^k f(t_i)[X(t_{i+1}) - X(t_i)] - Y |^2 \to 0 \tag{8.131}$$

as $k \to \infty$, for every sequence of partitions $0 = t_0 < t_1 < \ldots < t_k = T$ such that $\max(t_{i+1} - t_i) \to 0$. It is evident that the only difference with regard to the Stieltjes–Riemann integral known in the calculus of deterministic functions is in the specific understanding of the limit.

In order to construct a mathematically correct theory of stochastic differential equations, one has to introduce the integral of a stochastic process with respect to a stochastic process. This is the so-called Ito integral, which can be regarded as an extension of the deterministic Stieltjes–Lebesgue integral. It gives a formal basis for the general definition of a solution of stochastic differential equations, that is, differential equations fulfilled by stochastic processes. Such equations are called Ito differential equations.

We do not, however, have to go so deep into the difficult theory of stochastic processes. For our purposes it is sufficient to introduce a stationary white Gaussian noise U with zero mean and constant covariance matrix Z and the corresponding Wiener process W. The solution of equation (8.126) with initial condition $x(0)$ and the stochastic input U is a stochastic process defined by the formula

$$X(t) = \Phi(t)x(0) + \int_0^t \Phi(t - s) B \, dW(s). \tag{8.132}$$

In the sequel we shall deal with asymptotically stable systems. As time goes on, the solution (8.132) tends to the solution with zero initial condition

$$X(t) = \int_0^t \Phi(t - s) B \, dW(s), \tag{8.133}$$

that is, the influence of initial condition vanishes. Formula (8.133) may be used to define the solution of (8.126) 'after an infinite time from the beginning':

$$X(t) = \int_{-\infty}^{t} \Phi(t-s) B \, dW(s) \tag{8.134}$$

where the Wiener process W is defined on the whole real line.

We pass on to the evaluation of integral performance criterion on trajectories of system (8.126) driven by the stationary white Gaussian noise U. We assume zero initial conditions and take the stochastic process (8.133) as the solution of (8.126). Since the stochastic process $t \mapsto X(t)^T V X(t)$ is integrable, we may define the integral performance criterion in the form

$$J = \lim_{T \to \infty} E \frac{1}{T} \int_0^T X(t)^T V X(t) \, dt \tag{8.135}$$

where E is the operator of mean value and $V = V^T \geq 0$. This is the simplest and most natural adaptation of criterion (8.125) to systems with stochastic inputs. Under the assumption that all symbols have sense, we may change the order of integration and computing the mean, which gives

$$J = \lim_{T \to \infty} \frac{1}{T} \int_0^T E[X(t)^T V X(t)] \, dt. \tag{8.136}$$

The integrand can be computed with the use of the Wiener–Lee formulae (Korn and Korn, 1968, section 18.2), adapted to the multidimensional case. If we take into consideration that the input white Gaussian noise is stationary and system (8.126) is constant in time and asymptotically stable, we obtain

$$E[X(t)^T V X(t)] = \mathrm{tr}\, [Z B^T \int_0^t \Phi(s)^T V \Phi(s) \, ds \, B] \tag{8.137}$$

where tr denotes the trace of a matrix, that is, the sum of all elements on the diagonal. The reader who prefers considerations based on intuition to rigorous, but complicated proofs perhaps will be satisfied with the following simple derivation of formula (8.137) (Kwakernaak and Sivan, 1972, section 1.11).

$$X(t) = \int_0^t \Phi(t-s) B \, dW(s) = \int_0^t \Phi(t-s) B U(s) \, ds$$

$$= \int_0^t \Phi(s) B \bar{U}(s) \, ds \tag{8.138}$$

where $\bar{U}(s) = U(t-s)$. Then

$$E[X(t)^T V X(t)] = E \int_0^t \int_0^t \bar{U}(z)^T B^T \Phi(z)^T V \Phi(s) B \bar{U}(s) \, dz \, ds$$

$$= E \int_0^t \int_0^t \operatorname{tr}[\bar{U}(s)\bar{U}(z)^T B^T \Phi(z)^T V \Phi(s) B] \, dz \, ds$$

$$= \int_0^t \int_0^t \operatorname{tr}[E(\bar{U}(s)\bar{U}(z)^T) B^T \Phi(z)^T V \Phi(s) B] \, dz \, ds$$

$$= \int_0^t \int_0^t \operatorname{tr}[\delta(s-z) Z B^T \Phi(z)^T V \Phi(s) B] \, dz \, ds$$

$$= \operatorname{tr}\left[Z B^T \int_0^t \Phi(s)^T V \Phi(s) \, ds \, B \right]. \tag{8.139}$$

The lack of rigour arises in formula (8.138). After an integration by parts we obtain from (8.137)

$$\int_0^T E[X(t)^T V X(t)] \, dt = \operatorname{tr}\left[Z B^T \int_0^T (T-t) \Phi(t)^T V \Phi(t) \, dt \, B \right]. \tag{8.140}$$

Substituting this result into formula (8.135) we finally obtain

$$J = \operatorname{tr}\left[Z B^T \int_0^\infty \Phi(t)^T V \Phi(t) \, dt \, B \right]. \tag{8.141}$$

Note that this expression can be also obtained by virtue of the following equality:

$$\lim_{T \to \infty} \frac{1}{T} \int_0^T E[X(t)^T V X(t)] \, dt = \lim_{t \to \infty} E[X(t)^T V X(t)] \tag{8.142}$$

which may be verified by direct calculation. The performance criterion (8.141) can be expressed by means of the results of earlier chapters. For example, in the notation of section 5.1, we have

$$J = \operatorname{tr}(Z B^T M B). \tag{8.143}$$

The extension to systems with time delays is straightforward. For the system

Sec. 8.4] **Application to stochastic inputs** 195

$$\dot{x}(t) = \sum_{i=0}^{N} A_i x(t - h_i) + Bu(t), \quad t \geq 0 \\ x(t) \in \mathbb{R}^n, \; u(t) \in \mathbb{R}^m, \quad 0 = h_0 < h_1 < \ldots < h_N \Bigg\}$$ (8.144)

the results are formally identical, with a different definition of the fundamental matrix solution Φ which is given by

$$\dot{\Phi}(t) = \sum_{i=0}^{N} A_i \Phi(t - h_i)$$

$$\Phi(t) = 0, \quad t < 0, \; \Phi(0) = I.$$ (8.145)

The solution of (8.144) with zero initial conditions, driven by the stationary white Gaussian noise U, is given by (8.133) and the integral performance criterion (8.135) is expressed by formulae (8.141) and (8.143), with M defined in section 5.7.

Analogous expressions for the performance criterion may be easily obtained for the modifications of the problem, considered in earlier chapters (integrals with weighting functions and/or derivatives, time delays in the output or in the integrand of the criterion, etc.). However, in input-driven systems we encounter time delays in the control function. Consider the system equation

$$\dot{x}(t) = \sum_{i=0}^{N} [A_i x(t - h_i) + B_i u(t - h_i)], \quad t \geq 0 \\ x(t) \in \mathbb{R}^n, \; u(t) \in \mathbb{R}^m, \quad 0 = h_0 < h_1 < \ldots < h_N \Bigg\}.$$ (8.146)

Assume that the initial conditions are equal to zero and the input for positive t is the stationary white Gaussian noise $U(t)$ and is equal to zero for negative t. The solution of (8.146) then is a stochastic process determined by

$$X(t) = \int_0^t \Phi(t - s) \sum_{i=0}^{N} B_i \, dW(s - h_i)$$ (8.147)

where Φ is given by (8.145) and $W(s) = 0$ for $s \leq 0$. Formula (8.147) is equivalent to

$$X(t) = \sum_{i=0}^{N} \int_0^t \Phi(t - s - h_i) B_i \, dW(s).$$ (8.148)

Similarly as for the system without delays we compute $E[X(t)^T V X(t)]$ and finally obtain for the criterion (8.135)

$$J = \mathrm{tr} \, [Z \sum_{i,j=0}^{N} B_i^T \int_0^\infty \Phi(t - h_i)^T V \Phi(t - h_j) \, dt \, B_j]$$ (8.149)

or, using the notation of section 5.7,

$$J = \operatorname{tr}\left(Z \sum_{i,j=0}^{N} B_i^T L_{ij} B_j\right) \qquad (8.150)$$

where

$$L_{ij} = \begin{cases} L(h_j - h_i), & j \geq i \\ L(h_i - h_j)^T, & j < i \end{cases}. \qquad (8.151)$$

Let us now return to the control system of Example 5.1. Assume this time that its input is not the constant reference signal w but a stationary white Gaussian noise U with zero mean value and variance equal to Z. According to (5.16), the dynamics of the system is described by the equation

$$\dot{y}(t) = -Ky(t - h) + Ku(t) \qquad (8.152)$$

where u denotes a deterministic input. The output induced by the Gaussian noise U, with zero initial conditions, is therefore a stochastic process given by

$$Y(t) = \int_0^t \Phi(t - s)K\, dW(s). \qquad (8.153)$$

Φ is the solution of the initial value problem

$$\left.\begin{array}{l}\dot{\Phi}(t) = -K\Phi(t - h) \\ \Phi(0) = 1, \ \Phi(t) = 0 \text{ for } t < 0\end{array}\right\}, \qquad (8.154)$$

and W is the Wiener process corresponding to U. Let the integral performance criterion be

$$J = \lim_{T \to \infty} E \frac{1}{T} \int_0^T Y(t)^2\, dt. \qquad (8.155)$$

By virtue of the results of Chapter 5 (formula (5.53)) and formula (8.143), we immediately obtain

$$J = ZK^2 M$$
$$= ZK \frac{1 + \sin Kh}{2\cos Kh} = ZK \frac{\cos Kh}{2(1 - \sin Kh)}. \qquad (8.156)$$

8.5 CONVENTIONAL CONTROL SYSTEM DRIVEN BY WHITE GAUSSIAN NOISE

We shall now study the conventional control system of Fig. 8.2 driven by white Gaussian noise. We assume that the initial conditions are equal to zero, so is the reference input w, and the only non-zero inputs are the disturbances z_1 and z_2. To comply with the general theory given in section 8.4, we confine ourselves to PI-controllers. Similarly as in section 8.3 we assume that the delay is in the output of the controlled plant and so the system is described by the following set of equations (for $t \geqslant 0$):

$$u(t) = a_R \varepsilon(t) + b_R \int_0^t \varepsilon(\tau) \, d\tau \qquad (8.157)$$

$$\dot{x}(t) + qx(t) = K[u(t) + z_2(t)] \qquad (8.158)$$

$$y(t) = x(t - h) \qquad (8.159)$$

$$\varepsilon(t) = z_1(t) - y(t). \qquad (8.160)$$

If we denote $x_1 = x$ and define a new variable

$$x_2(t) = \int_0^t \varepsilon(\tau) \, d\tau$$

we may rewrite the system equations (8.157)–(8.160) in the form

$$\begin{pmatrix} \dot{x}_1(t) \\ \dot{x}_2(t) \end{pmatrix} = \begin{pmatrix} -q & \beta \\ 0 & 0 \end{pmatrix} \begin{pmatrix} x_1(t) \\ x_2(t) \end{pmatrix} + \begin{pmatrix} -\alpha & 0 \\ -1 & 0 \end{pmatrix} \begin{pmatrix} x_1(t-h) \\ x_2(t-h) \end{pmatrix} + \begin{pmatrix} \alpha & K \\ 1 & 0 \end{pmatrix} \begin{pmatrix} z_1(t) \\ z_2(t) \end{pmatrix}. \qquad (8.161)$$

The sense of equation (8.161) driven by white Gaussian noise z_1, z_2 is explained in section 8.4. We denote

$$B = \begin{pmatrix} \alpha & K \\ 1 & 0 \end{pmatrix}. \qquad (8.162)$$

In our problem it is more sensible to take the squared output $y(t)^2$ as the integrand in the performance criterion J so that J might help us to evaluate damping, or attenuation properties of the control system with respect to stochastic noise. We thus define

$$J = \lim_{T \to \infty} \frac{1}{T} \int_0^T E[y(t)^2] \, dt = \lim_{T \to \infty} \frac{1}{T} \int_0^T E[x_1(t)^2] \, dt. \qquad (8.163)$$

If we now put

$$V = \begin{pmatrix} 1 & 0 \\ 0 & 0 \end{pmatrix},$$

we obtain by virtue of (8.150)

$$J = \text{tr}\,[ZB^T L(0) B] \qquad (8.164)$$

where $L(0)$ is determined in section 5.7 by solving the Lyapunov system of equations. Instead of computing $L(0)$ along those lines we shall here use the alternative method based on Parseval's theorem. To this end we consider system (8.161) with $z_1 = z_2 = 0$ and $x_1(t) = x_2(t) = 0$ for $t < 0$. The values of $x_1(0)$ and $x_2(0)$ are arbitrary. The integral

$$J_1 = \int_0^\infty x_1(t)^2 \, dt = \int_0^\infty x(t)^T V x(t) \, dt \tag{8.165}$$

can be written in the following form:

$$J_1 = \begin{pmatrix} x_1(0) \\ x_2(0) \end{pmatrix}^T L(0) \begin{pmatrix} x_1(0) \\ x_2(0) \end{pmatrix} \tag{8.166}$$

by virtue of section 5.7. On the other hand, J_1 can be easily calculated with the use of Parseval's theorem method. From equation (8.161), we obtain the Laplace transform of x_1,

$$\hat{x}_1(s) = \frac{s x_1(0) + \beta x_2(0)}{s(s+q) + (\alpha s + \beta) e^{-sh}}. \tag{8.167}$$

In the case where $\beta \neq 0$ we obtain

$$J_1 = T_1[\beta^2 x_2(0)^2 - \rho^2 x_1(0)^2] + T_2[\beta^2 x_2(0)^2 + \sigma^2 x_1(0)^2] \tag{8.168}$$

where

$$\rho = \sqrt{\frac{\Delta + q^2 - \alpha^2}{2}}, \quad \sigma = \sqrt{\frac{\Delta - q^2 + \alpha^2}{2}} \tag{8.169}$$

$$\Delta = \sqrt{(q^2 - \alpha^2)^2 + 4\beta^2}, \tag{8.170}$$

$$T_1 = \frac{\beta - \alpha q - (\rho^2 - q^2) \cosh \rho h}{\alpha \rho^2 - q\beta + \rho(\rho^2 - q^2) \sinh \rho h} \cdot \frac{1}{2\Delta} \tag{8.171}$$

$$T_2 = \frac{\beta - \alpha q + (\sigma^2 + q^2) \cos \sigma h}{\alpha \sigma^2 + q\beta - \sigma(\sigma^2 + q^2) \sin \sigma h} \cdot \frac{1}{2\Delta}. \tag{8.172}$$

Comparing this result with (8.166) we find out that the (2×2) matrix $L(0)$ has the form

$$L(0) = T_1 \begin{pmatrix} -\rho^2 & 0 \\ 0 & \beta^2 \end{pmatrix} + T_2 \begin{pmatrix} \sigma^2 & 0 \\ 0 & \beta^2 \end{pmatrix}. \tag{8.173}$$

Denoting

$$Z = \begin{pmatrix} Z_1 & Z_{12} \\ Z_{12} & Z_2 \end{pmatrix} \tag{8.174}$$

we finally calculate

$$J = T_1 \operatorname{tr} \left[ZB^T \begin{pmatrix} -\rho^2 & 0 \\ 0 & \beta^2 \end{pmatrix} B \right] + T_2 \operatorname{tr} \left[ZB^T \begin{pmatrix} \sigma^2 & 0 \\ 0 & \beta^2 \end{pmatrix} B \right]$$

$$= T_1[Z_1(\beta^2 - \alpha^2 \rho^2) - 2Z_{12} K \alpha \rho^2 - Z_2 K^2 \rho^2]$$

$$+ T_2[Z_1(\beta^2 + \alpha^2 \sigma^2) + 2Z_{12} K \alpha \sigma^2 + Z_2 K^2 \sigma^2]. \tag{8.175}$$

Notice that if $\beta = \alpha q$, then T_1 is indeterminate. Its value is determined by passing to the limit $\beta \to \alpha q$.

If $\beta = 0$, the Laplace transform of x_1 is given by

$$\hat{x}_1(s) = \frac{x_1(0)}{s + q + \alpha e^{-sh}}. \tag{8.176}$$

From a comparison with (8.75) it evidently follows that:

$$J_1 = \frac{x_1(0)^2}{2\tau} \frac{q\alpha + (\tau^2 - q^2)\cosh \tau h}{\alpha\tau + (\tau^2 - q^2)\sinh \tau h} \quad \text{if } \alpha^2 < q^2, \tag{8.177}$$

$$J_1 = \frac{x_1(0)^2}{2\tau} \frac{q\alpha - (\tau^2 + q^2)\cos \tau h}{-\alpha\tau + (\tau^2 + q^2)\sin \tau h} \quad \text{if } \alpha^2 > q^2, \tag{8.178}$$

and

$$J_1 = \frac{x_1(0)^2}{4q}(1 + qh) \quad \text{if } \alpha = q \neq 0, \tag{8.179}$$

where

$$\tau = \sqrt{|\alpha^2 - q^2|}.$$

Finally, we obtain

$$J = T(\alpha^2 Z_1 + 2\alpha K Z_{12} + K^2 Z_2) \tag{8.180}$$

where

$$T = \frac{1}{2\tau} \frac{q\alpha + (\tau^2 - q^2)\cosh \tau h}{\alpha\tau + (\tau^2 - q^2)\sinh \tau h} \quad \text{if } \alpha^2 < q^2, \tag{8.181}$$

$$T = \frac{1}{2\tau} \frac{q\alpha - (\tau^2 + q^2)\cos \tau h}{-\alpha\tau + (\tau^2 + q^2)\sin \tau h} \quad \text{if } \alpha^2 > q^2, \tag{8.182}$$

and

$$T = \frac{1 + qh}{4q} \quad \text{if } \alpha = q \neq 0. \tag{8.183}$$

To obtain a clear, qualitative picture, let us confine the analysis of optimal settings to two extreme cases: (1) $Z_1 = Z_{12} = 0$, $Z_2 > 0$ and (2) $Z_1 > 0$, $Z_2 = Z_{12} = 0$. In the first case, the performance functional J (8.175) or (8.180) differs from that given by formulae (8.74) or, respectively, (8.77)–(8.79) by a constant positive factor, equal to $Z_2 K^2$. This fact has a simple physical interpretation. Functional (8.163) in the system driven by white Gaussian noise z_2 has the same value as functional (8.52) (with $\gamma = 0$) in the system activated by initial conditions $x(0) = K\sqrt{Z_2}$, $x(t) = 0$ for $t < 0$. In consequence, the analysis of optimal settings done in section 8.3 applies also in this case.

In the second case, the optimal solution is trivial: $a_R = 0$ for the P-controller, and $a_R = b_R = 0$ for the PI-controller. In cases intermediate between (1) and (2), the optimal settings are also intermediate. Generally, the optimal values of a_R are smaller than in case (1).

9

Predictive control: mismatch problems

9.1 INTRODUCTION

In this chapter we shall apply methods met earlier to problems which arise in the predictive control of a system with a single series delay. The effects of the delay on stability, when there is feedback control, are of course evident from the earlier chapters, but in addition to the stability effects of the delay there are further problems associated with the control which are, in part, common to all control schemes, whether the given system is delay-free or not. It is important in all control problems that the system to be controlled is well understood, and hence well modelled, so that we may represent it by a transfer function. It is the dual purpose of control to change the dynamics of the given system for the better, and to be able to do this even if the system is not as well modelled as one would wish.

If the system to be controlled has a series delay term we shall need to know the value of delay for control design, and for stability studies, the more accurately the better.

A plant with delay-free part, $G(s)$, and series delay, h, represented by $G(s)\exp(-sh)$, could be controlled by delay-free methods if a series controller with transfer function $\exp(+sh)$ could be produced to 'cancel' the delay term. The transfer function $\exp(+sh)$ represents a pure predictor which is of course not physically realizable. Predictive control methods are techniques which attempt, in various ways, to overcome this difficulty.

In as much as the accuracy of prediction will depend on how well modelled the plant is, especially the delay, it is important to know how sensitive the performance is, however quantified, to differences between the *actual* parameters of the plant and what the parameters are believed to be, i.e. the nominal values, when the controller is designed. The effects of these differences between a plant and its model are called **mismatch** problems. **Temporal mismatch** is the difference between the actual delay, and its nominal value, that assumed for design, based on one's best knowledge of the delay.

Parametric mismatch denotes the corresponding difference between the delay-free part of the plant and its model. It is convenient to use subscripts for the nominal 'model' values. Hence $G_0(s)$ and $\exp(-h_0 s)$ denote the transfer function models of $G(s)$ and $\exp(-sh)$. h_0 is the **model delay**, $G_0(s)$ the **sub-plant model**.

The predictor scheme which serves as an example for this chapter is that due to O.J.M. Smith, the most quoted, if not the best understood, method of control for systems with a series delay. What follows is sufficient for this chapter to be complete but the following references will be found helpful: Smith (1957), Marshall (1979), and Morari and Zafiriou (1989). The last-mentioned gives a particularly well-informed and balanced discussion of the controversies concerning the use of Smith's method. It is hoped that the rest of the chapter, with its new insights into the mismatch problem in particular, will keep the debate more well informed than hitherto.

9.2 A PREDICTOR CONTROL SCHEME

Consider the system in Fig. 9.1. If the point A is available for control (i.e. is accessible) then the control problem reduces trivially to a delay-free design, as the delay may

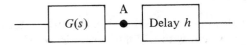

Fig. 9.1. System with output delay.

be excluded from the control loop. We shall assume (realistically) that A is inaccessible. The signal at A resulting from a known input stimulus may be inferred by using a model, $G_0(s)$, of $G(s)$, and the output from this may be used for control purposes. Let us denote by $C(s)$ the series controller for the unity-negative-feedback control of $G_0(s)$. The resulting control, Fig. 9.2, is clearly open loop, and there will be difficulties especially if $G(s)$ is open-loop unstable.

This naive open-loop control of $G(s)$, may be improved upon in several ways. One will need at least to be able to do something about poor identification, i.e. the given $G(s)$ is not known with precision, so that $G(s)$ and $G_0(s)$ differ significantly. The output $y_p(t)$ from $G_0(s)$ is clearly, if $G_0(s) = G(s)$, a prediction of the output from the delay, i.e. $y_p(t) = y(t + h)$.

It is important to be clear what the conditions need to be for such prediction.
(i) $G_0(s) = G(s)$.
(ii) Input $x(t)$ is known for all $t \geqslant 0$.
(iii) There are no disturbances.
(iv) There are no (unknown) initial conditions on $G(s)$, due to previous inputs.
(v)   There is no function stored in the delay, i.e. $y(t) = 0$ for $t \leqslant h$.

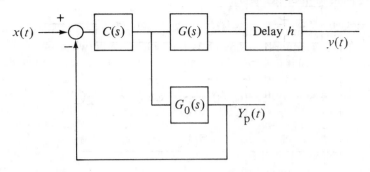

Fig. 9.2. Open-loop control.

If we assume that **all** these conditions are met then we may make use of a principle due to O.J.M. Smith (1957). This control method, resulting from Smith's principle, is applicable only to systems of the series type, with delay in input or output, or both. There is no feedback delay here. We shall describe the delayed output case.

Assume that a controller $C(s)$ can be designed for the unity-negative-feedback control of the delay-free, **known**, plant $G(s)$. For a particular input, $x(t)$, chosen from that class for which the $C(s)$ is appropriate, denote the corresponding output by $y(t)$. In practice $C(s)$ may well be a PI-controller; we assume it may be represented by a delay-free transfer function, $C(s)$.

For the case where $G(s)$ has an added series delay, h, with the input to the delay being inaccessible, as discussed earlier, Smith's principle asserts that $C(s)$ should be replaced by a controller $C^*(s)$ so that the output from the delay is a delayed version, by h, of the corresponding $y(t)$ of the delay-free design. In transfer function terms the principle requires that

$$X(s) \frac{C^*(s)G(s)\exp(-sh)}{1 + C^*(s)G(s)\exp(-sh)} = X(s) \frac{C(s)G(s)\exp(-sh)}{1 + C(s)G(s)}. \tag{9.1}$$

Hence

$$C^*(s) = \frac{C(s)}{1 + C(s)G(s)(1 - \exp(-sh))}, \tag{9.2}$$

which results in the control scheme shown as Fig. 9.3. Its relation to Fig. 9.2 is obvious: we have prediction of output together with an additional outer feedback loop.

There are equivalent and important rearrangements of this figure, but this realization emphasizes the prediction. It is particularly important to recall that the $G_0(s)$ and the $\exp(-sh_0)$ added for control purposes are models of the sub-plant and the series delay. Either because of inaccuracies of simulation or because of identification we shall need to distinguish the actual sub-plant from their models, and we use this notation introduced earlier. The transfer function (input–output) of the system of Fig. 9.3 is

204 Predictive control: mismatch problems   [Ch. 9

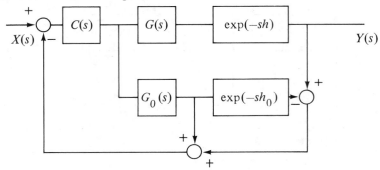

Fig. 9.3. Predictor control scheme of O.J.M. Smith.

$$\frac{Y(s)}{X(s)} = \frac{C(s)G(s)\exp(-sh)}{1 + C(s)G_0(s) + C(s)(G(s)\exp(-sh) - G_0(s)\exp(-sh_0))}. \quad (9.3)$$

Under the assumptions listed, i.e. assuming perfect matching where $h = h_0$ and $G_0(s) = G(s)$, the transfer function reduces to

$$\frac{C(s)G(s)\exp(-sh)}{1 + C(s)G(s)},$$

as it should, by Smith's principle.

9.3 MISMATCH

When $G_0(s) = G(s)$, but $h_0 \neq h$, we have temporal mismatch and the denominator of (9.3), which determines the stability of the control scheme, gives rise to the characteristic equation:

$$1 + CG + CG(z - z_0) = 0, \quad (9.4)$$

where $z = \exp(-sh)$, $z_0 = \exp(-sh_0)$.

Similarly for the parametric mismatch case $h_0 = h$, and $G_0(s) \neq G(s)$, and the characteristic equation is now

$$1 + CG + C(G - G_0)(z - 1) = 0. \quad (9.5)$$

Recall that $1 + CG = 0$ is the characteristic equation when there is no mismatch. Hence it is evident that, when there is temporal mismatch, the magnitude of the CG term is important, and conversely when there is parametric mismatch that the term $(1 - z)$ is important. We note this to be small, when $h(= h_0)$ is small (Garland and Marshall, 1974). For parametric mismatch we note that the corresponding stability problem and the related performance problem are single-delay problems. However, in the temporal mismatch case ($h_0 \neq h$) the corresponding stability problem is complicated unless the h_0 and h are commensurate, in which case the methods of Chapter 2 may be applied directly. We give examples in the next chapter of stability for the non-commensurate case, in the context of predictor schemes.

9.4 MISMATCH AND PERFORMANCE

In the temporal mismatch case we use the notation $J(h, h_0) = \int_0^\infty (x(t) - y(t))^2 \, dt$ for the cost, when delay and its model take values h and h_0 respectively. A similar notation will be used in the parametric, and mixed cases.

When there is matching, so that Smith's principle is satisfied, the output $y(t)$ is a delayed version of the delay-free output, i.e. $y(t) = w(t - h)$, where h is the delay and $w(t)$ is the response of the delay-free system. Hence in the Smith Predictor case, when the input is a step function $H(t)$

$$J(h, h) = \int_0^\infty (x(t) - y(t))^2 \, dt = \int_0^\infty (x(t) - w(t - h))^2 \, dt$$

$$= \int_0^h (x(t) - w(t - h))^2 \, dt + \int_h^\infty (x(t) - w(t - h))^2 \, dt$$

$$= \int_0^h x^2(t) \, dt + \int_h^\infty (x(t - h) - w(t - h))^2 \, dt \qquad (9.6)$$

as $x(t) = x(t - h)$ for $t > h$.

Hence

$$J(h, h) = \int_0^h x^2(t) \, dt + \int_0^\infty (x(t) - w(t))^2 \, dt$$

$$= h + \int_0^\infty (x(t) - w(t))^2 \, dt. \qquad (9.7)$$

The integral term is the *delay-free cost*.

The h term arises from the absence of output for $0 < t < h$, and hence is an inevitable (and irreducible) error which arises in the control of any system with series delay, *whatever the control scheme*.

We shall define as the 'Smith prediction cost' $J_{SP} = \int_0^\infty (x(t) - y(t))^2 \, dt - h$. At matching this will be equal, as we have just shown, to the delay-free cost. This is the value of cost that Smith's principle, strictly applied (i.e. $G_0 = G$, $h_0 = h$), will give. Recall that $C(s)$ was chosen, assuming $G(s)$ and h known, to minimize the delay-free cost, so that *at matching* this is the minimum cost for the *given* structure of $C(s)$, i.e. this is the best that can be achieved *at matching*, if $C(s)$ is chosen via parametric optimization for a given controller structure. So that at matching, changing $C(s)$ will give poorer performance. Recall that the design method of a Smith predictor is a delay-free method, i.e. $C(s)$ is designed by delay-free methods, and the controller design is completed by adding the models of sub-plant and delay, assumed accurately known. In examples to follow we shall explore the effects of mismatch on $J_{SP} = \int_0^\infty (x(t) - y(t))^2 \, dt - h$. We shall explore (perversely?) the effects of regarding $G_0(s)$ and h_0 as design parameters, to see what the effects on J_{SP} might be. Further, when $h_0 \neq h$, and/or $G_0 \neq G$, we could also explore the effects of modifying $C(s)$. Note in particular at matching, J_{SP} is constant, and independent of h. Of interest (as will be seen shortly) is the possibility of reduction of J_{SP} by mismatch, which though counter-intuitive from a control point of view is to give fresh insights into SP control.

The following example, of extreme mismatch in delay only, demonstrates an improvement (i.e. a smaller J) by deliberate mismatch of delay. In this example h_0 is replaced by 0, so that $\exp(-sh_0)$ is unity, and we have an example of extreme temporal mismatch.

Example 9.1
Let $C(s) = K$, $G(s) = 1/s$, $G_0(s) = 1/s$, $X(s) = 1/s$ and let $h_0 = 0 \ (\neq h)$. Hence

$$Y(s) = \frac{1}{s} \frac{K \exp(-sh)}{s + K + K(\exp(-sh) - 1)}$$

and

$$E(s) = \frac{1}{s + K \exp(-sh)}$$

and

$$J(h, 0) = \frac{1}{2K} \frac{\cos Kh}{(1 - \sin Kh)},$$

as shown earlier. Hence

$$J_{SP}(h, 0) = \frac{1}{2K} \frac{\cos Kh}{(1 - \sin Kh)} - h. \tag{9.8}$$

In the matched case this should take the delay-free value,

$$J_{SP}(h, h) = \int_0^\infty (x(t) - w(t))^2 \, dt = \frac{1}{2\pi i} \int_{-i\infty}^{+i\infty} \frac{1}{s + K} \frac{1}{-s + K} \, ds = \frac{1}{2K}. \tag{9.9}$$

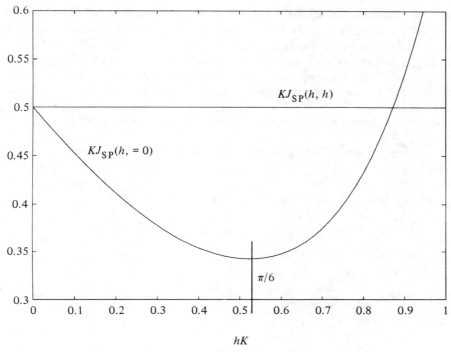

Fig. 9.4. $J_{SP}(h, 0)$ showing the minimum.

The graph of $J_{SP}(h, 0)$ (see Fig. 9.4) shows that J_{SP} is less than $1/(2K)$ for a range of Kh values. Note also the (expected) instability at $hK = \pi/2$ shown in Fig. 9.5. The minimum value (where $\partial J_{SP}/\partial h = 0$) is at $hK = \pi/6$ (∼0.52) and the improvement over the Smith (matched) case is about 31%.

Having demonstrated that mismatch in this case of delay improved performance, which is contrary to the spirit of Smith's principle, it is important to show that this is not an isolated example. Further, this suggests that the performance of some delay-free systems could be improved by the use of controllers with delay as opposed to delay-free controllers (Marshall and Salehi, 1982).

9.5 SENSITIVITY AND THE SIGN OF MISMATCH: THE TEMPORAL CASE

By Taylor's expansion of $J(h, h_0)$ about the matched value, assuming that mismatch is in delay only, we have, at $h_0 = h + \Delta h$,

$$J(h, h + \Delta h) = J(h, h) + \frac{\partial J}{\partial h_0}(h, h)\Delta h$$

$$+ \frac{\partial^2 J}{\partial h_0^2}(h, h)\frac{(\Delta h)^2}{2} + \text{higher-order terms.} \qquad (9.10)$$

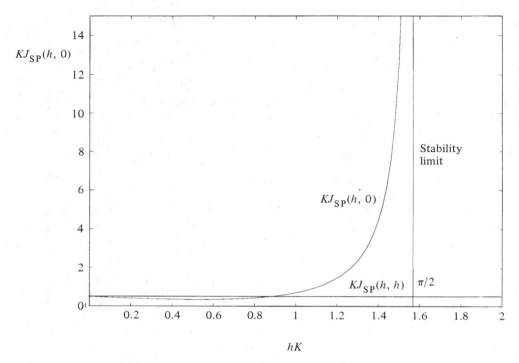

Fig. 9.5. $J_{SP}(h, 0)$ showing instability.

Improvement in performance will be obtained if $J(h, h + \Delta h) - J(h, h) < 0$. For small mismatch this implies that $\Delta h\, \partial J(h, h)/\partial h_0 < 0$ implies an improvement in performance. Hence the sign of mismatch for improvement depends on the sign of the temporal partial derivative, to be called the *temporal sensitivity coefficient*. If the sensitivity coefficient is positive then negative values of Δh improve performance; this is called 'under-estimation of delay'. Conversely 'over-estimation', $\Delta h > 0$, will improve performance if the sensitivity coefficient is negative at matching. In general the sign of the coefficient will depend on h. We now derive an expression for $\partial J(h, h)/\partial h_0$ for the general case.

For an SP scheme with mismatch in delay only we have

$$E(s, h, h_0) = X(s)\left(1 - \frac{CGz}{1 + CG + CG(z - z_0)}\right),$$

where

$$z = \exp(-sh), \quad z_0 = \exp(-sh_0).$$

Hence

Sec. 9.5] Sensitivity and the sign of mismatch: the temporal case

$$J(h, h_0) = \frac{1}{2\pi i} \int_{-i\infty}^{+i\infty} E(s, h, h_0)E(-s, h, h_0)\,ds = \frac{1}{2\pi i} \int_{-i\infty}^{+i\infty} E\bar{E}\,ds, \qquad (9.11)$$

say. It is convenient to introduce the notation $\bar{z}(s) = \exp(+sh)$, and $\bar{z}_0(s) = \exp(+sh_0)$. Differentiating we have

$$\frac{\partial J(h, h_0)}{\partial h_0} = \frac{1}{2\pi i}\int_{-i\infty}^{+i\infty}\frac{\partial}{\partial h_0}(E\bar{E})\,ds = \frac{1}{2\pi i}\int_{-i\infty}^{+i\infty}\bar{E}\frac{\partial E}{\partial h_0}\,ds + \frac{1}{2\pi i}\int_{-i\infty}^{+i\infty}E\frac{\partial \bar{E}}{\partial h_0}\,ds.$$

By symmetry as elsewhere (replacing s by $-s$ in the second integral) it follows that

$$\frac{\partial J(h, h_0)}{\partial h_0} = \frac{2}{2\pi i}\int_{-i\infty}^{+i\infty}\bar{E}\frac{\partial E}{\partial h_0}\,ds,$$

$$\frac{\partial E}{\partial h_0} = \frac{X(s)sz_0(CG)^2 z}{(1 + CG + CG(z - z_0))^2}$$

and

$$\bar{E} = X(-s)\left(1 - \frac{\bar{C}\bar{G}\bar{z}}{1 + \bar{C}\bar{G} + \bar{C}\bar{G}(\bar{z} - \bar{z}_0)}\right).$$

Hence at matching the integral becomes

$$\left.\frac{\partial J}{\partial h_0}\right|_{h_0=h} = \frac{2}{2\pi i}\int_{-i\infty}^{+i\infty}\frac{X(s)X(-s)s(CG)^2 z^2 (1 + \bar{C}\bar{G} - \bar{C}\bar{G}\bar{z})}{(1 + \bar{C}\bar{G})(1 + CG)^2}\,ds. \qquad (9.12)$$

Closing the integral to the right (due to the z numerator term) leads to an evaluation over the roots of $1 + \bar{C}\bar{G} = 0$ only. Hence this is a delay-free calculation; the explicit numerator exponential term introduces no new difficulty.

$$\therefore \left.\frac{\partial J}{\partial h_0}\right|_{h_0=h} = -2\sum_{1+\bar{C}\bar{G}=0}\operatorname{res}\frac{X(s)X(-s)s(CG)^2 z}{(1 + \bar{C}\bar{G})(1 + CG)^2}. \qquad (9.13)$$

For the case $CG = K/s$, and $X(s) = 1/s$, it is easily shown by substitution in this formula that $(\partial J/\partial h_0)_{h_0=h} = \tfrac{1}{2}\exp(-Kh)$, which is sign-definite (positive), so that near the matched condition, for any $h > 0$, a value of h_0 less than h will improve performance. Knowledge of the first derivative only is insufficient to find an optimum value of Δh_0 but the sign of mismatch, which is an important design consideration, may be found readily by a delay-free calculation.

A second example is the following: $CG = K^2/[s(s + K)]$, which is an optimized delay-free transfer function. Here,

$$\left(\frac{\partial J}{\partial h_0}\right)_{h_0=h} = -\frac{1}{\sqrt{3}}e^{-\frac{Kh}{2}}\sin\left(\frac{\sqrt{3}Kh}{2}\right),$$

which is a straightforward residue calculation, at two complex conjugate roots.

Note this expression is not sign-definite, being negative for $0 < \sqrt{3}Kh/2 < \pi$ and alternating in sign over intervals of π. For Kh taking values in the first interval, i.e. $Kh < 2\pi/\sqrt{3}$, a positive value of offset—over-estimation—improves performance. It is known elsewhere (Byron, Cox and Ball, 1979) that over-estimation improves performance for second-order sub-plants. This example shows that over-estimation is not always appropriate in all second-order cases, but is h dependent. Clearly, owing to the oscillating magnitude of $(\partial J/\partial h_0)_{h_0 = h}$ the same offset will result in different values of improvement. There will be more improvement at some values of h than others. h is, however, not a parameter that one can change at will. It is possible to exploit this dependence on h, when h is small, by deliberate addition of delay in the outer loop of the SP control scheme. This technique is discussed in Chotai (1980), Hocken, Salehi and Marshall (1983), in the context of simulation studies.

9.6 THE PARAMETRIC SENSITIVITY COEFFICIENT

Let us consider a sub-plant $G = G(s, \alpha)$ where α is one of the sub-plant parameters. Denote the corresponding sub-plant model by $G_0 = G(s, \alpha_0)$. As the mismatch is assumed to be in the sub-plant only, we have $z = z_0 = \exp(-sh)$. Hence we have for erorr,

$$E(s) = X(s)\left(1 - \frac{CGz}{1 + CG_0 + Cz(G - G_0)}\right).$$

By methods very similar to the temporal case we obtain

$$\left.\frac{\partial J}{\partial \alpha_0}\right|_{\alpha_0 = \alpha} = \frac{2}{2\pi i} \int_{-i\infty}^{+i\infty} \frac{X(s)X(-s)\bar{C}^2\bar{G}(1-\bar{z})}{(1+\bar{C}\bar{G})^2}\left[\bar{z} - \frac{CG}{1+CG}\right]\left.\frac{\partial \bar{G}_0}{\partial \alpha_0}\right|_{\alpha_0 = \alpha} ds \quad (9.14)$$

and the contour must be closed to the left, owing to the $\bar{z} = \exp(+sh)$ term.

Residues will arise from $1 + CG = 0$, as singularities due to $\partial \bar{G}_0/\partial \alpha_0$ will lie in the RH plane if the sub-plant is open-loop stable, which it is in the context of SP control.

When $CG = K^2/[s(s + K)]$ and $CG_0 = K^2/[s(s + a)]$ we obtain, when $h = h_0$, and $X(s) = 1/s$,

$$\left(\frac{\partial J}{\partial a}\right)_{G_0 = G} = \frac{-2}{2\pi i}\int_{-i\infty}^{+i\infty} \frac{K^4(1 - \exp(sh))((s^2 + Ks + K^2)\exp(sh) - K^2)}{s^2(s-K)(s^2 + Ks + K^2)(s^2 - Ks + K^2)^2} ds.$$

(9.15)

With closure to the left, and noting that there is no singularity at $s = 0$, the resulting evaluation of residues, at the LH conjugate roots of $s^2 + sK + K^2 = 0$, results in the expression

$$\left(\frac{\partial J}{\partial a}\right)_{a = K} = \frac{1}{3K^2}\left[1 - \exp\left(\frac{-Kh}{2}\right)\cos\left(\frac{\sqrt{3}Kh}{2}\right)\right]. \quad (9.16)$$

9.7 CROSS-SENSITIVITY

We have spoken so far about the effect of offset in either delay, or a parameter in the delay-free part. We now consider briefly the sensitivity with respect to h_0, near the temporal matched value, when there is a fixed offset (necessarily small) in a parameter. The mixed derivative $\partial^2 J/\partial h_0 \partial \alpha_0$ now plays an important rôle, for if $\alpha_0 = \alpha + \Delta\alpha$, say, we have

$$\frac{\partial}{\partial h_0} J(h, h_0, \alpha, \alpha_0)\bigg|_{\substack{h_0=h \\ \alpha_0 \neq \alpha}} = \frac{\partial}{\partial h_0} J(h, h_0, \alpha, \alpha)\bigg|_{h_0=h}$$

$$+ \Delta\alpha \frac{\partial^2}{\partial h_0 \partial \alpha_0} J(h, h_0, \alpha, \alpha_0)\bigg|_{\substack{h_0=h \\ \alpha_0=\alpha}}$$

$$+ \text{higher-order terms}. \qquad (9.17)$$

This is zero at $h = 0$, and sign definite (positive) for positive values of Kh. Hence under-estimation of the parameter a improves performance. The degree of improvement is clearly h dependent, and increasing, broadly, with h. For large values of h the magnitude of improvement might be very small—and importantly the effect on stability of mismatch would need careful consideration.

The sign of the temporal sensitivity now depends on the sign of the parametric offset, and its magnitude, and the sign and magnitude of the mixed derivative. For the case $CG_0 = K^2/(s(s + a))$, with $a = K$ at matching, it is possible to show that

$$\left(\frac{\partial^2 J}{\partial h_0 \partial a}\right)_{\substack{h_0=h \\ a=K}} = \frac{1}{K}\left\{\frac{Kh\beta}{6}(\cos\theta + \sqrt{3}\sin\theta)\right.$$

$$+ \frac{2\sqrt{3}}{9}\beta\sin\theta + \beta^2\left(\frac{\cos 2\theta}{3}\right)$$

$$\left. - \frac{2}{9}(1-\beta^2) - \frac{3}{2}(1-\beta\cos\theta) - \frac{1}{6}\beta\cos\theta\right\} \qquad (9.18)$$

where $\beta = \exp[-Kh/2]$ and $\theta = (\sqrt{3}/2)Kh$.

At $h = 2$, $a = K = \frac{1}{2}$ we find that

$$\frac{\partial J}{\partial h_0}\bigg|_{\substack{h_0=h \\ a=K}} = -\frac{1}{\sqrt{3}} e^{\frac{-hK}{2}} \sin\frac{\sqrt{3}Kh}{2} = -0.2667 \qquad (9.19)$$

and

$$\frac{\partial^2 J}{\partial h_0 \partial a}\bigg|_{\substack{h_0=h \\ a=K}} = -1.5915$$

so that the mixed derivative is rather important here, and if $\Delta\alpha$ is negative the sensitivity changes sign in the region of $\Delta\alpha = -0.167$, so that if α is in error by about 30% of its nominal value the temporal sensitivity changes sign.

9.8 HIGHER DERIVATIVES

We summarize in Table 9.1 the second derivatives corresponding to some given CG. The second derivatives are obtained by the obvious extension to those for the first derivatives. For example, the second temporal sensitivity is given by

$$\left.\frac{\partial^2 J}{\partial h_0^2}\right|_{h_0=h} = -\frac{1}{\pi i}\int_{-i\infty}^{+i\infty} X(s)X(-s)s^2(\bar{C}\bar{G})^2$$

$$\times \left\{\frac{(CG)^2}{(1+CG)^2(1+\bar{C}\bar{G})^2} + \bar{z}\frac{(1+\bar{C}\bar{G}+2\bar{C}\bar{G}\bar{z})}{(1+CG)(1+\bar{C}\bar{G})^3}\right\} ds \qquad (9.20)$$

in general.

Table 9.1. Temporal sensitivity coefficients.

$CG = K/s$	$\left.\dfrac{\partial J}{\partial h_0}\right\|_{h_0=h} = \dfrac{1}{2}\exp(-Kh)$
	$\left.\dfrac{\partial^2 J}{\partial h_0^2}\right\|_{h_0=h} = \dfrac{K}{2}\Big(1-\exp(-Kh)-\exp(-2Kh)\Big)$
$CG = \dfrac{K^2}{s(s+K)}$	$\left.\dfrac{\partial J}{\partial h_0}\right\|_{h_0=h} = \dfrac{-1}{\sqrt{3}}\exp(-hK)\sin\left(\dfrac{\sqrt{3}hK}{2}\right)$
	$\left.\dfrac{\partial^2 J}{\partial h_0^2}\right\|_{h_0=h} = K\left(1+\dfrac{1}{\sqrt{3}}\exp(-Kh)\sin(\sqrt{3}Kh)\right)$ $+ \dfrac{1}{2\sqrt{3}}\exp\left(\dfrac{-Kh}{2}\right)\sin\left(\dfrac{\sqrt{3}Kh}{2}\right)$ $- \dfrac{1}{2}\exp\left(\dfrac{-Kh}{2}\right)\cos\left(\dfrac{\sqrt{3}Kh}{2}\right)$

For the *parametric case* $CG = K^2/[s(s+K)]$, $CG_0 = K^2/[s(s+a)]$ we have

$$\left(\frac{\partial J}{\partial a}\right)_{a=K} = \frac{1}{3K^2}\left(1 - \exp\left(-\frac{Kh}{2}\right)\cos\left(\frac{\sqrt{3}hK}{2}\right)\right) \qquad (9.21)$$

and

$$\left.\frac{\partial^2 J}{\partial a^2}\right|_{a=K} = \frac{1}{9K^3}(18hK - 2(1-\beta^2) - 5(1-\beta\cos\theta) - 11\sqrt{3}\beta\sin\theta$$

$$-\sqrt{3}\beta^2\sin 2\theta + hK\beta(3\cos\theta - \sqrt{3}\sin\theta)) \tag{9.22}$$

where $\beta = \exp(-hK/2)$ and $\theta = hK\sqrt{3}/2$.

9.9 SUMMARY

In this chapter we have discussed the effects of mismatch in either the delay or the sub-plant, or briefly both, for a predictor scheme, that of O.J.M. Smith. Similar analysis can be applied to other predictor schemes, of course, including extensions of the SP scheme (Marshall, 1974, Gawthorp, 1977, Morari and Zafiriou, 1989). That the performance of predictor schemes may be **improved** by mismatch is true for cases where the controller *structure* is decided *a priori*. This will usually be the case when the corresponding delay-free design is via optimization of a controller with prescribed structure, i.e. by 'analytic design' (Jacobs, 1974, Newton, Gould and Kaiser, 1957, Åström, 1970). If the delay-free controller has been found using optimal control methods where the controller structure is found by synthesis, i.e. by optimal LQP methods for example, then there will be no improvement by mismatch (Hocken and Marshall, 1982, 1983, Hocken, Marshall and Salehi, 1983).

That is not to say that the methods of this chapter are inappropriate in the 'optimal' case. The assumption made (rather grossly) for many optimal control schemes, with delay or not, is that the system to be controlled is modelled exactly. There is still a mismatch problem, and the corresponding sensitivities of cost with respect to plant (and controller) parameters are very important. It is surely realistic to apply mismatch ideas to all control problems when the design is based on the notion of a 'given' system or an 'internal model'. The ideas of this chapter have not been overtaken by robust control, where plant uncertainty is postulated *a priori*. In real problems, some parameters will be well known, and some others will have error bounds. It will be helpful in the case of the control of time-delay systems, where we have seen intuition can be a poor guide, to make as much as possible of the deterministic parts, and to explore the uncertainties by way of sensitivity analysis and the methods of this chapter. This is to argue that there is a spectrum of approaches, from simulation, where a variety of special cases may be considered, to the analytical methods we give here which are part of engineering science.

In the chapter to follow we explore further cases of mismatch with closed-form solutions, where possible, of the corresponding costs, and further discussion of the related stability problems.

10

Predictive control: exact cost functionals

10.1 INTRODUCTION

The sensitivity methods discussed in the previous chapter enabled discussion of mismatch problems in which the mismatch, temporal or parametric, was small, with the sign of differential coefficients being sufficient to indicate how cost functionals would change in the region of the matched cases. It is possible to calculate approximate values of mismatch for which improvement would be best by truncation of the Taylor series about the matched values, and using the given expressions for first and second derivatives. The accuracy of this is of course difficult to check, so that assessment of 'best' mismatch and the degree of improvement over the matched case can be no more than well-informed estimates. These values will depend very much on the magnitude of the delay, and the magnitude of the 'gain' of the delay-free part.

In this chapter we derive exact cost functional results for cases of mismatch. The parametric case, where the delays of plant and model are strictly equal, reduces to a single-delay problem for which exact solutions are possible using the methods met earlier.

Another case where exact solutions are possible are those for which the delay h_0 is set to zero. This is an example of extreme temporal mismatch. How extreme depends on the value of plant delay. There are some counter-intuitive results here, especially for a first-order plant, that are not repeated in any general way for plants of higher order. This will enable us to draw conclusions about the use of the Smith predictor for first-order systems.

We discuss, with examples, stability of temporal mismatch cases, showing how to find the minimum value of mismatch at which instability occurs and the corresponding delay. Similarly where there is parametric mismatch only, we find the range of mismatch for which stability will not occur, whatever the delay.

A further class of problems for which exact solutions are possible is that where the delay model $\exp(-sh_0)$ is replaced in the predictor scheme by an approximant with the use of an all-pass Padé model, using the properties met in Chapter 7. This is a technique sometimes advocated as a practical method of time-delay control. This problem is a single-delay problem for which exact solutions may be found, as the sum of a finite number of residues.

Further problems for which exact cost expressions may be found are those mismatch problems where the delay and the model delay are related by integers enabling the commensurate methods of Chapter 4 to be applied.

It is instructive to compare the results of these commensurate calculations with the corresponding results where a Padé model is used in the predictor.

The problems, both of stability and of cost functional evaluation for predictive control, will be problems with two delays, not necessarily equal. When the delays are equal, or one of them is zero (i.e. $h_0 = 0$), the problem reduces to a single-delay problem, for which a very direct technique involving symmetries is given in section 10.6. Exact results, as shown in Chapter 4, are possible when the delays are commensurate, for example $h_0 = h/2$, or $2h$. In general for **cost functional** evaluation as closed-form solutions in terms of a finite sum of terms will not be possible. The corresponding stability problems reduce to a type of single-delay problem, which reduces, as section 10.2 shows, to a straightforward numerical problem.

Each of the examples will be shown in sufficient detail for the methods to be followed completely. Routine algebra will not be given fully, except where use is made of helpful algebraic identities that have been used for purposes of simplification.

We have remarked, especially in Chapters 1 and 2, upon the relationships between cost function and stability. Usually the denominators of the cost functional expressions that are found will vanish for values of h for which the characteristic equation has roots on the imaginary axis. Hence we would expect a close relationship between such denominators and the real or imaginary parts of the equation satisfied by the poles of the system, or of the error. This may be used in two ways, either to check the consistency of the denominator with these real and imaginary parts, or to be exploited in simplifying the cost functional or its evaluation. This latter will be used to enable us to derive cost functionals more efficiently in the examples. Other algebraic techniques will be used here and there to simplify the expressions obtained for cost functionals, in a systematic way.

More sensitivity results will be given in this chapter, in particular for the case where h_0 is set equal to zero. This will enable us to comment further on the appropriateness of the Smith predictor method with respect to the magnitude of the plant delay.

10.2 PARAMETRIC MISMATCH: STABILITY

We know that if a delay-free system is stable, then the corresponding matched Smith predictor scheme is also stable. In this section we shall explore the effect on stability of mismatch in the delay-free part of the time-delay, assuming that the delay is modelled exactly. This is closely related to the problem 'stability independent of delay' in single-delay problems. We are in fact looking at stability independent of the common delay $h\ (=h_0)$.

Recall from Chapter 2 that if a system is stable at $h = 0$ and if $W(\omega^2) > 0$ for all $\omega > 0$ then there will be no purely imaginary roots of the characteristic equation, whatever the value of delay.

Example 10.1

We first look at the problem $CG = K/s$, $CG_0 = K'/s$, $h_0 = h$, where K, K' are positive.

For an SP scheme the characteristic equation is

$$1 + CG_0 + CGz - CG_0 z_0 = 0. \tag{10.1}$$

Using $z_0 = z$ and substituting, we have the characteristic equation $s + K' + (K - K')z = 0$, so that

$$A(s) = s + K' \quad \text{and} \quad C(s) = (K - K').$$

Hence

$$A\bar{A} - C\bar{C} = K'^2 - s^2 - (K - K')^2 \tag{10.2}$$

and $W(\omega^2) = \omega^2 + 2KK' - K^2$.

This has a positive slope w.r.t. ω^2 so that a necessary and sufficient condition for stabillity independent of delay is that $2K' - K > 0$. Hence the system is stable for all values of the (common) delay when $K' > K/2$. This clearly includes $K' = K$, at matching. In terms of parametric mismatch, $(K' - K) > -K/2$.

Example 10.2

In this example we find the range of values of a for which the predictor system with $CG = K^2/[s(s + K)]$ and $CG_0 = K^2/[s(s + a)]$ is stable independent of the common delay $h\ (=h_0)$. The characteristic equation is:

$$1 + \frac{K^2}{s(s+a)} + z\left(\frac{K^2}{s(s+K)} - \frac{K^2}{s(s+a)}\right) = 0, \tag{10.3}$$

which after substituting $s = Kp$, $a = KA$ gives $p(p + 1)(p + A) + (p + 1) + (A - 1)z = 0$, and $W_A = \omega^6 + \omega^4(A^2 - 1) + \omega^2(A^2 - 1) + A(2 - A)$. The derivative (w.r.t. ω^2) will be useful and is $W'_A = 3\omega^4 + 2\omega^2(A^2 - 1) + (A^2 - 1)$.

It is convenient to replace ω^2 by x. We need to find the range of values of A such that $W_A > 0$ for $x > 0$. It is necessary that $W_A(0) = A(2 - A) > 0$. Note that at matching, $A = 1$ and $W_1 = x^3 + 1 > 0$ (as expected), and further for $A > 1$, $W'_A > 0$ for all $x \geqslant 0$, so that there will be stability independent of delay when $1 \leqslant A < 2$, at least. For values of $A < 1$: we note that at $A = 0$, $W_0 = x^3 - x^2 - x = x(x^2 - x - 1)$, which has roots at 0, and $x = (1 \pm \sqrt{5})/2$, so that this (noting that $W'_0(0) = -1$) implies a value A_C satisfying $0 < A_C < 1$ for which there will be the limiting case of $W_{A_C}(x)$ touching the x-axis, from above.

The values of x and A which satisfy this satisfy $x^3 + (a^2 - 1)x^2 + (a^2 - 1)x + A(2 - a) = 0$ and $3x^2 + 2(A^2 - 1)x + (A^2 - 1) = 0$, simultaneously. Note that the first equation may be written as a quadratic in A, i.e.

$$A^2(x^2 + x - 1) + 2A + x(x^2 - x - 1) = 0.$$

A simple iterative scheme using these two quadratics, seeking values of A satisfying $0 < A < 1$, and $x > 0$, was found to converge very rapidly to the solution $A = 0.427088\ldots$ at $x = 0.861433\ldots$. In terms of the original problem we find stability independent of the common delay is obtained for $0.427088 < a/K < 2$. Hence broadly speaking the SP scheme is stable when the delays are matched (at whatever value), and for a two-fold increase, or decrease, of the mismatch parameter a.

10.3 PARAMETRIC MISMATCH: COST

When plant delay and the model delay are equal then the corresponding parametric mismatch problem is a single-delay problem, so that exact results are possible for cost function values when the parameters of the sub-plant and its model are not equal. We consider the example $CG = K/s$, $CG_0 = K'/s$, where K, K' are positive numbers. In this case, when $x(t) = H(t)$

$$E(s) = X(s) - Y(s)$$
$$= \frac{1}{s} \frac{(s + K'(1 - z))}{s + K' + (K - K')z}$$

and

$$J = -\frac{1}{2\pi i} \int_{-i\infty}^{+i\infty} \frac{1}{s^2} \frac{(s + K'(1 - z))(-s + K'(1 - \bar{z}))}{(s + K' + (K - K')z)(-s + K' + (K - K')\bar{z})} ds. \quad (10.4)$$

This is a single-delay calculation. Note that $sE(s) \to 0$ as $s \to 0$ so that there are no poles at $s = 0$. In this case the value of J becomes, after the usual manipulation for a single-delay problem:

$$J = \sum_{s=0, \pm \lambda} \text{res} \frac{s + K' - K'z}{s^2(s + K' + (K - K')z)} \frac{s^2 - KK'}{s^2 - \lambda^2} \quad (10.5)$$

where $\lambda^2 = 2KK' - K^2$.

Note that λ is real when $K' > K/2$.

Evaluating the residues gives

$$J = \frac{K'(1 + hK')}{\lambda^2} - \frac{K^2}{2\lambda^3} \left\{ \frac{(K - K') \sinh \lambda h + \lambda}{(K - K') \cosh \lambda h + K'} \right\} \quad \text{for } K' > K/2 \quad (10.6a)$$

and

$$J = -\frac{K'(1 + hK')}{\mu^2} + \frac{K^2}{2\mu^3} \left\{ \frac{(K - K') \sin \mu h + \mu}{(K - K') \cos \mu h + K'} \right\} \quad \text{for } K' < K/2. \quad (10.6b)$$

where $\lambda > 0$ and satisfies $\lambda^2 = 2KK' - K^2$ and $\mu > 0$ satisfies $\mu^2 = K^2 - 2KK'$.

In the special case of $K' = K$, $\lambda = K$ and $J = 1/(2K) + h$. For this matched case we have $J_{SP} = J - h = 1/(2K)$, the delay-free cost value, as it should.

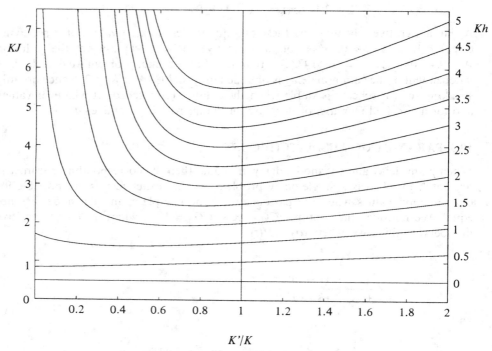

Fig. 10.1. KJ vs K'/K for various delays.

Figs 10.1 and 10.2 show the values of KJ and KJ_{SP}, respectively, plotted against K'/K. Note that there is no discontinuity at $K' = K/2$.

For large values of delay, minima lie close to the matched value. It is noted that improvement in performance is small. Fig. 10.2 shows most strongly the effect of mismatch on performance. If improvement is defined as $100(J(\text{matched}) - J(\text{min}))/J(\text{matched})$ we find that the greatest improvement is at $Kh \approx 0.6$ where a 13.8% improvement is to be found. This reduces to 0.3% at $Kh = 5$, at which value of Kh there is little point in deliberate mismatch. There is never improvement if K' exceeds K (Walton and Marshall, 1984b).

Further single-delay calculations for which $J(h, 0)$ have been found follow in section 10.6.

10.4 TEMPORAL MISMATCH: STABILITY

A Smith predictor scheme is stable for all $h_0 = h$ when there is matching of the sub-plant ($G = G_0$). However, there may well be instability when $G = G_0$, if h and its model h_0 differ, i.e. under temporal mismatch. It is important to know what the minimum value of $|h_0 - h|$ should be for instability to occur, and to know the corresponding value(s) of h.

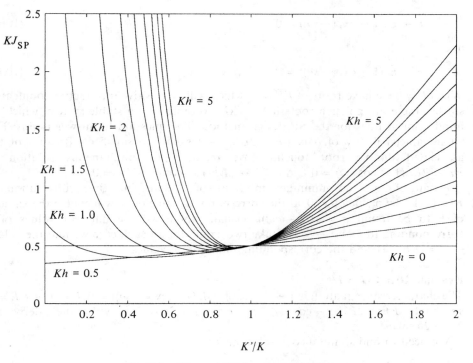

Fig. 10.2. KJ_{SP} vs K'/K for various delays.

It is convenient to write $h_0 = h + \delta$, where δ may take positive or negative values. For temporal mismatch, $G = G_0$ so that the characteristic equation is

$$1 + CG + CG(z - z_0) = 0, \tag{10.7}$$

i.e.

$$1 + CG + CGz(1 - \exp(-s\delta)) = 0. \tag{10.8}$$

If CG is strictly proper then so is $CG/(1 + CG)$, and we may write $CG/(1 + CG) = E/F$ where the polynomial F is of higher order than the polynomial E.

Write the characteristic equation as

$$F + Ez(1 - \exp(-s\delta)) = 0 \tag{10.9}$$

and by the methods used in Chapter 2 we may eliminate z for purely imaginary values of s.

Writing $F(-s) = \bar{F}$, as earlier, we find that

$$F\bar{F} - E\bar{E}(1 - \exp(-s\delta))(1 - \exp(+s\delta))|_{s=i\omega} = 0 \tag{10.10}$$

$$= F\bar{F} - E\bar{E}(2 - 2\cos\omega\delta) = 0 \tag{10.11}$$

is satisfied for common purely imaginary roots of

$$F + Ez(1 - \exp(-s\delta)) = 0 \tag{10.12}$$

and

$$\bar{F} + \bar{E}\bar{z}(1 - \exp(+s\delta)) = 0. \tag{10.13}$$

When $\delta = 0$ we have roots of $F\bar{F} = 0$, which will be positive (product of conjugates) so that there are no such roots if $1 + CG = 0$ represents a stable system, which by implication it does for the SP design method. The question to answer is this: For what value(s) if any of δ is $F\bar{F} - 2E\bar{E}(1 - \cos\omega\delta) = 0$ satisfied? Because of the associated double root touching we require the simultaneous solution of $F\bar{F} - 2E\bar{E}(1 - \cos\omega\delta) = 0$ and $d(F\bar{F} - 2E\bar{E}(1 - \cos\omega\delta))/d\omega = 0$.

Having found the minimum modulus of δ satisfying this, substitution in $F + Ez(1 - \exp(-i\omega\delta)) = 0$, at the corresponding value of ω, will enable the value(s) of h to be found ($z = \exp(-i\omega h)$). Usually there will be different values of h corresponding to δ and $-\delta$. We take two examples. In each case we find the values of $\delta = |h_0 - h|$, and the corresponding minimum h values.

Example 10.3 $CG = 1/s$
The characteristic equation is $1 + s + \exp(-sh)(1 - \exp(-s\delta)) = 0$. $F = 1 + s$, $E = 1$ so that $F\bar{F} - 2E\bar{E}(1 - \cos\omega\delta)$ is $\omega^2 - 1 + 2\cos\omega\delta$, with the derivative $2\omega - 2\delta\sin\omega\delta$.

We need to find simultaneous solutions of

$$\left.\begin{array}{l}\omega^2 - 1 + 2\cos\omega\delta = 0 \\ \omega - \delta\sin\omega\delta = 0,\end{array}\right\} \tag{10.14}$$

i.e. of $\omega\delta\sin\omega\delta = 1 - 2\cos\omega\delta = \omega^2$. This gives $\omega = 1.3483$, $|\delta| = 1.4775$.

Now $\exp(-sh) = -(1 + s)/(1 - \exp(-s\delta))$ at $s = i\omega$ and $\delta = \pm 1.4775$. Denote by h_+ and h_- the minimum positive values of h satisfying this equation for the two values of δ. Note that the right-hand side of the equation should have unity modulus, at the values of ω and δ obtained earlier, and this serves as a useful check.

It is easily shown that h_+ and $h_- = \pi/\omega - \delta$, and the values are $h_- = 1.2120$, and $h_+ = 2.0646$. If CG in this example is replaced by K/s the corresponding values of δ, h_- and h_+ are decreased by the factor K, and ω similarly increased: Now $h_0 = h + \delta$, so that, when δ is positive, $h_0 = h_+ + \delta = 2.0646 + 1.4775 = 3.5421$. When δ is negative, $h_0 = 1.2102 - 1.4775 + 2\pi q/\omega$ to give $h_0 = 4.3914$, at $h = 5.8689$ and $q = 1$. The corresponding $q = 0$ case gives rise to negative h_0, which we dismiss.

Hence we conclude in this example that the minimum value of $|h_0 - h|$, for which there is instability, occurs at $\delta = +1.4775$, at which $h(=h_+) = 2.0646$, $h_0 + 3.5421$, and the system is stable for $0 < |\delta| < 1.4775$ at all values of h.

Example 10.4 $CG = CG_0 = K^2/s(s + K)$
The characteristic equation is

$$s(s + K) + K^2 + K^2 z(1 - \exp(-s\delta)) = 0 \tag{10.15}$$

and the equations to be simultaneously satisfied are

$$p^4 - p^2 - 1 + 2\cos p\Delta = 0 \\ p(1 - 2p^2) + \Delta \sin p\Delta = 0 \Bigg\} \tag{10.16}$$

where the dimensionless variables $p = \omega/K$ and $\Delta = K\delta$ have been introduced for convenience.

Solving these equations is equivalent after elimination of the trig. terms to finding values of Δ, p to satisfy

$$\Delta^2 = 4p^2(1 - 2p^2)^2/(4 - (p^4 - p^2 - 1)) \tag{10.17}$$

$$\Delta = \frac{1}{p} \cos^{-1}((1 + p^2 - p^4)/2). \tag{10.18}$$

Values of p, Δ are found to be 0.97801 and 1.04597 respectively.

Now

$$-z = \frac{1 - p^2 + ip}{1 - \cos p\Delta + i \sin p\Delta} = -\exp(-iKph)$$

from which it follows that

$$pKh_+ = 2q\pi - \frac{3\pi}{2} - \frac{p\Delta}{2} - \tan^{-1}\left(\frac{p}{1 - p^2}\right) \tag{10.19}$$

and

$$pKh_- = 2q\pi - \frac{3\pi}{2} + \frac{p\Delta}{2} - \tan^{-1}\left(\frac{p}{1 - p^2}\right) \tag{10.20}$$

at $p = 0.9780$, to give 3.78064 for the minimum positive value of Kh_+, and 5.85612 for the minimum positive value of kh_-. Hence minimum values of $|h_0 - h|$ occur for $h_0 = h_+ + \delta = 3.78064/K$ when $\delta = 1.04597/K$ and we conclude that provided the mismatch in delay is less than $1.04597/K$ then the system is stable for all values of delay.

10.5 TEMPORAL MISMATCH: $h_0 = 0$

In the previous chapter it was found helpful to know the sensitivity of a cost, at matching, with respect to h_0. It is similarly helpful to know the sensitivity when $h_0 = 0$. This calculation is possible for an SP scheme because with h_0 set equal to zero we have a single-delay problem, the only delay being the 'given' plant delay. It is convenient to represent the derivative as $\partial J(h, 0)/\partial h_0$, without ambiguity.

Now $J = \int_{-i\infty}^{i\infty} E(s)E(-s) \, ds/2\pi i$ where E depends on s, and on z and z_0. h_0 appears in the z_0 term only and $\partial z_0/\partial h_0 = -sz_0 = -s$ at $h_0 = 0$.

$$\frac{\partial J}{\partial h_0} = \frac{1}{2\pi i} \int_{-i\infty}^{+i\infty} \frac{\partial}{\partial h_0} E(s)E(-s) \, ds$$

$$= \frac{1}{2\pi i} \int_{-i\infty}^{+i\infty} \left(E(s) \frac{\partial E(-s)}{\partial h_0} + E(-s) \frac{\partial E(s)}{\partial h_0} \right) ds \qquad (10.21)$$

so that, exploiting symmetry, we have

$$\frac{\partial}{\partial h_0} J(h, 0) = \frac{1}{\pi i} \int_{-i\infty}^{+i\infty} E \frac{\partial E(-s)}{\partial h_0} ds. \qquad (10.22)$$

With step input we have, defining error in the usual way,

$$E(s) = \frac{1}{s} \left(1 - \frac{CGz}{1 + CG_0 + CGz - CG_0 z_0} \right). \qquad (10.23)$$

so that

$$E(s)|_{h_0 = 0} = \frac{1}{s} \left(1 - \frac{CGz}{1 + CGz} \right) = \frac{1}{s} \left[\frac{1}{1 + CGz} \right] = \frac{1}{s + Kz}$$

when $CG = K/s$.

Note that this is independent of G_0, so that this result is true if $G_0(s) \neq G(s)$. We also note that this is input minus output for the case of delayed unity negative feedback of the plant $G(s)$ with the usual delay-free controller $C(s)$. We shall exploit this later in this section when calculating costs.

$\partial \bar{E}/\partial h_0$ is equal to

$$\frac{\partial}{\partial h_0} \left(\frac{1}{-s} \left(1 - \frac{\bar{C}\bar{G}}{1 + \bar{C}\bar{G}_0 + \bar{C}\bar{G}\bar{z} - \bar{C}\bar{G}\bar{z}_0} \right) \right)$$

$$= \frac{1}{-s} \frac{\bar{C}\bar{G}(-\bar{C}\bar{G}_0)(s\bar{z}_0)\bar{z}}{(1 + \bar{C}\bar{G}_0 + \bar{C}\bar{C}\bar{z} - \bar{C}\bar{G}_0\bar{z}_0)^2}. \qquad (10.24)$$

At matching of G and G_0, and with $h_0 = 0$, this gives

$$\left. \frac{\partial \bar{E}}{\partial h_0} \right|_{h_0=0} = \frac{(\bar{C}\bar{G})^2 \bar{z}}{(1 + \bar{C}\bar{G}\bar{z})^2} = \frac{K^2 \bar{z}}{(s - K\bar{z})^2} \qquad (10.25)$$

when $CG = K/s$. Hence

$$\frac{\partial J(h, 0)}{\partial h_0} = \frac{1}{\pi i} \int_{-i\infty}^{i\infty} \frac{1}{s + Kz} \frac{K^2 \bar{z}}{(s - K\bar{z})^2} ds$$

$$= 2 \sum_{s + Kz = 0} \text{res} \frac{K^2 \bar{z}}{(s + Kz)(s - K\bar{z})^2}$$

$$= -2 \sum_{s + Kz = 0} \text{res} \frac{K^3 s}{(s + Kz)(s^2 + K^2)^2} \qquad (10.26)$$

since $\bar{z} = -K/s$, when $s + Kz = 0$.

As the sum of all residues is zero

$$\frac{\partial J(h,0)}{\partial h_0} = +2 \sum_{s=\pm iK} \text{res} \frac{K^3 s}{(s+Kz)(s^2+K^2)^2}$$

$$= 4K^3 \, \text{Re} \, \frac{d}{ds}\left(\frac{s}{(s+Kz)(s+iK)^2}\right)_{s=iK}. \qquad (10.27)$$

Performing the differentiation and substituting $s = iK$ we obtain, noting that $z = \exp(-iKh)$, $\bar{z} = \exp(iKh)$:

$$\frac{\partial J(h,0)}{\partial h_0} = \frac{\text{Re}(\exp(iKh) - Kh)}{2(1 - \sin Kh)} = \frac{\cos Kh - Kh}{2(1 - \sin Kh)}. \qquad (10.28)$$

Hence $\partial J(h,0)/\partial h_0 = 0$ when $\cos hK = hK$, i.e. $hK = 0.74\ldots$; also $\partial J(h,0)/\partial h_0$ is positive for $hK < 0.74\ldots$. This fact together with the value of $J(h,0)$, which is easily evaluated, as we shall shortly show, has important implications for the applicability (or otherwise!) of the SP scheme when $Kh < 0.74$. Now

$$E(s)|_{h_0=0} = \frac{1}{s+Kz},$$

so that

$$J = \frac{1}{2\pi i} \int_{-i\infty}^{i\infty} \frac{1}{s+Kz} \frac{1}{-s+K\bar{z}} \, ds.$$

We have shown in Chapter 4 that for this value of $E(s)$ the corresponding cost is $\cos Kh/[2K(1 - \sin Kh)]$, so that the unmatched SP cost here is

$$J_{SP} = \frac{\cos Kh}{2K(1 - \sin Kh)} - h. \qquad (10.29)$$

Recall that the matched SP cost, $J_{SP}(h,h) = 1/2K$. Hence we may conclude that for the control of the system $CGz = [K \exp(-sh)]/s$ the mismatched cost is *less than the matched cost* when $hK < 0.74$ and is a minimum (at $\partial J_{SP}(h,0)/\partial h = 0$) when $hK = \pi/6$, and the cost there is

$$\frac{1}{2K}\left(\frac{\cos hK}{1 - \sin hK} - hK\right)_{hK=\pi/6} = 0.68483\ldots/(2K). \qquad (10.30)$$

This is about 31.5% better than the corresponding matched SP value!

Because $\partial J(h,0)/\partial h_0$ is positive for $hK < 0.74$ it is clear that in this case of controlling an integrator plus delay provided that $hK < 0.74$ the best strategy is to use negative feedback rather than the SP scheme! We shall later consider what to do if h exceeds the value $0.74/K$.

This example is included for its couter-intuitive property, i.e. poor mismatch (of the worst kind) improving a result. We would be very surprised if there were not some new applications of this. It is of course well known (Marshall, 1979, Bateman, 1945) that delay sometimes improved performance of a **delay-free** system. Here we see that delay improves the performance of unity negative feedback round an integrator, i.e. make K as large as possible and then add the feedback delay $\pi/6K$ to improve the performance by 31.5%. The practicality of the choice of the maximum value of K we leave to the practitioner.

We now give two second-order examples arising from $CG = K^2/[s(s + a)]$, and the special case when $a = K$, and the second-order system $CG = K(s + a)/s^2$.

For these examples we exploit further the simplifcation that the use of symmetry brings. This enables us to find a quicker route to the evaluation of J when $\Delta(s) = A\bar{A} - C\bar{C}$ has purely real and/or purely imaginary roots in s.

10.6 EVALUATION OF J WHEN $D(s) = 0$ and $h_0 = 0$

Recall the general form of error

$$E(s) = \frac{B(s) + D(s)\exp(-sh)}{A(s) + C(s)\exp(-sh)}$$

met in Chapters 2 and 4, and denote $\exp(-sh)$ by z.

We shall exploit the identity, valid for $\Delta(s) = 0$,

$$\frac{1}{A + Cz} = \frac{1}{C}\frac{(C - \bar{A}\bar{z})}{(A - \bar{A}) + (Cz - \bar{C}\bar{z})}. \tag{10.31}$$

This identity, with denominator the odd (or imaginary) part of $A + Cz$, leads to powerful, and relatively quick calculation of the single-delay integrals, reducing the evaluation, as we shall show, to finding real and imaginary (even and odd) parts of expressions, together with some straightforward 'partial fraction' techniques. We shall need to find the roots of $\Delta(s) = A\bar{A} - C\bar{C}$, as before.

Evaluation of J at a pair of real roots, and at a pair of purely imaginary roots: Let the positive root of $\Delta(s)$ be α and write $\Delta(s) = (s^2 - \alpha^2)\Delta_\alpha(s)$. Then using the quoted identity, and with $D = 0$,

$$J = - \sum_{W=0} \text{res} \frac{AB\bar{B}(C - \bar{A}\bar{z})}{(s^2 - \alpha^2)\Delta_\alpha(s)C(A - \bar{A} + Cz - \bar{C}\bar{z})}. \tag{10.32}$$

Denote by J_α the contribution of the residues at $s = \pm\alpha$

$$J_\alpha = - \sum_{\pm\alpha} \text{res} \frac{B\bar{B}(A - \bar{C}\bar{z})}{(s^2 - \alpha^2)\Delta_\alpha(s)(A - \bar{A} + Cz - \bar{C}\bar{z})} \tag{10.33}$$

using $A\bar{A} - C\bar{C} = 0$ and

$$J_\alpha = -\frac{(B\bar{B})(A - \bar{C}\bar{z})}{2\alpha\Delta_\alpha(\alpha)(A - \bar{A} + Cz - \bar{C}\bar{z})}\bigg|_{s=\alpha}$$

$$-\frac{(B\bar{B})(A - \bar{C}\bar{z})}{(-2\alpha)\Delta_\alpha(\alpha)(A - \bar{A} + Cz - \bar{C}\bar{z})}\bigg|_{s=-\alpha}. \tag{10.34}$$

But $(B\bar{B})$ is even in s, and $(A - \bar{A} + Cz - \bar{C}\bar{z})$ is odd in s. Hence

$$J_\alpha = -\frac{(B\bar{B})[(A - \bar{C}\bar{z}) + (\bar{A} - Cz)]}{2\alpha\Delta_\alpha(\alpha)(A - \bar{A} + Cz - \bar{C}\bar{z})}\bigg|_{s=\alpha}$$

$$= -\frac{(B\bar{B})\,\mathrm{Ev}(A - \bar{C}\bar{z})}{2\alpha\Delta_\alpha(\alpha)\,\mathrm{Odd}(A + Cz)}\bigg|_{s=\alpha} \tag{10.35}$$

with a similar expression for all other real roots.

For the contribution J_β from the purely imaginary pair, $\pm i\beta$, and writing $\Delta(s) = (s^2 + \beta^2)\Delta_\beta(s)$, we obtain, similarly,

$$J_\beta = \frac{(B\bar{B})\,\mathrm{Re}(A - \bar{C}\bar{z})}{2\beta\Delta_\beta(i\beta)\,\mathrm{Im}(A + Cz)}\bigg|_{s=i\beta} \tag{10.36}$$

We shall use these expressions to derive the costs for some second-order examples of extreme mismatch in delay. They are applicable to any *single*-delay problem for which $E(s)$ takes the form $B/(A + Cz)$.

Example 10.5 $CG = K^2/(s(s+a))$

$$E = \frac{1}{s}\left\{1 - \frac{CGz}{1 + CG_0 + CG(z - z_0)}\right\} = \frac{1}{s}\left\{\frac{1}{1 + CGz}\right\}. \tag{10.37}$$

Hence $E = (s+a)/[s(s+a) + K^2 z]$. Hence $\Delta(s) = -s^2(a^2 - s^2) - K^2 = s^4 - a^2 s^2 - K^4$ with roots $s^2 = (a^2 \pm (a^4 + 4K^4)^{1/2})/2$ so that $s = \pm\alpha, \pm i\beta$, where $\alpha^2, \beta^2 = ((a^4 + 4K^4)^{1/2} \pm a^2)/2$. Now $A = s(s+a)$, $B = s + a$, $C = K^2$, $D = 0$. At $s = \alpha$, $\mathrm{Ev}(A - \bar{C}\bar{z}) = \alpha^2 - K^2 \cosh\alpha h$, and $\mathrm{Odd}(A + Cz) = a\alpha - K^2 \sinh\alpha h$. At $s = i\beta$, $\mathrm{Re}(A - \bar{C}\bar{z}) = -\beta^2 - K^2 \cos\beta h$, and $\mathrm{Im}(A + Cz) = a\beta - K^2 \sin\beta h$. Hence

$$J(h, 0) = -\frac{(a^2 - \alpha^2)(\alpha^2 - K^2 \cosh\alpha h)}{2\alpha(\alpha^2 + \beta^2)(a\alpha - K^2 \sinh\alpha h)}$$

$$+ \frac{(a^2 + \beta^2)(+\beta^2 + K^2 \cos\beta h)}{2\beta(\alpha^2 + \beta^2)(a\beta - K^2 \sin\beta h)}. \tag{10.38}$$

Now $s^4 - a^2 s^2 - K^4 = (s^2 - \alpha^2)(s^2 + \beta^2) = s^4 - (\alpha^2 - \beta^2)s^2 - \alpha^2\beta^2$ and $\alpha^2\beta^2 = K^4$, $\alpha^2 - \beta^2 = a^2$ so that

$$J(h, 0) = \frac{\alpha^2(\beta^2 + K^2 \cos\beta h)}{2\beta(\alpha^2 + \beta^2)(a\beta - K^2 \sin\beta h)} + \frac{\beta^2(\alpha^2 - K^2 \cosh\alpha h)}{2\alpha(\alpha^2 + \beta^2)(a\alpha - K^2 \sinh\alpha h)}. \tag{10.39}$$

At $h = 0$ this expression reduces, using the relationships between α, β, K and a, to $(K^2 + a^2)/2aK^2$, which agrees with the delay-free calculation as a check.

The result for $a = K$ may be found by inspection, and the corresponding delay-free value of $1/K$ is found.

Fig. 10.3 shows the corresponding $J_{\mathrm{SP}} = J(h, 0) - h$. We note that the cost exceeds the Smith cost except at $Kh = 0$ only.

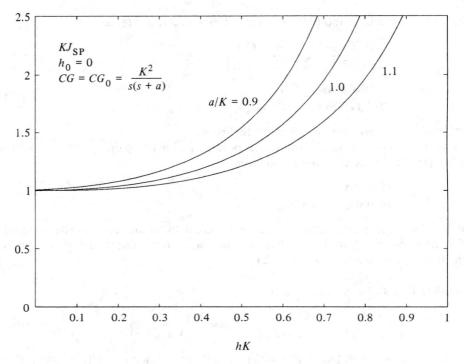

Fig. 10.3. $KJ_{SP}(hK, 0)$ vs hK for various a/K: $CG = CG_0 = K^2/(s(s + a))$.

Example 10.6
When

$$CG = \frac{K(s + a)}{s^2}, \quad A = s^2, \quad B = s, \quad C = K(s + a), \quad D = 0,$$

and

$$\Delta(s) = s^4 + K^2 s^2 - a^2 K^2 = (s^2 - \alpha^2)(s^2 + \beta^2),$$

where α^2, β^2 are given by $\alpha^2, \beta^2 = (\sqrt{K^4 + 4a^2 K^2} \mp K^2)/2$ and by a similar method to that of Example 10.3.

$$J(h, 0) = \frac{\alpha(\alpha^2 - Ka \cosh \alpha h + K\alpha \sinh \alpha h)}{2(\alpha^2 + \beta^2)(K\alpha \cosh \alpha h - Ka \sinh \alpha h)}$$
$$+ \frac{\beta(\beta^2 + Ka \cos \beta h - K\beta \sin \beta h)}{2(\alpha^2 + \beta^2)(K\beta \cos \beta h - Ka \sin \beta h)}. \qquad (10.40)$$

This simplifies by multiplying numerator and denominator of the α-expression by $Ka \sin(\alpha h) + K\alpha \cosh(\alpha h)$, and the β-expression similarly by $Ka \sin(\beta h) + K\beta \cos(\beta h)$, respectively, to obtain, after some routine algebra,

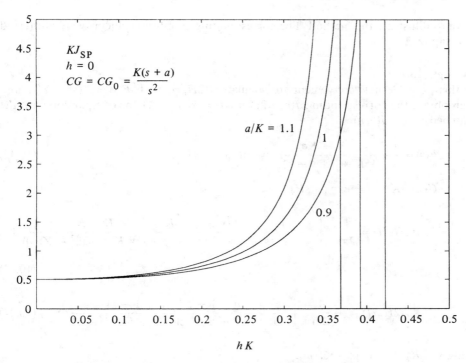

Fig. 10.4. $KJ_{SP}(hK, 0)$ vs hK for various a/K: $CG = CG_0 = K(s + a)/s^2$.

$$J = \frac{1}{2(\alpha^2 + \beta^2)} \left\{ \frac{aK + \beta^2 \cos \beta h}{K - \beta \sin \beta h} - \frac{aK - \alpha^2 \cosh(\alpha h)}{K + \alpha \sinh(\alpha h)} \right\}. \tag{10.41}$$

From this must be subtracted h to give the corresponding J_{SP}. This gives agreement with the delay-free value of $1/2K$.

Fig. 10.4 shows the corresponding $J_{SP} = J(h, 0) - h$, for various values of a. Note the onset of instability in this unmatched case for modest values of hK. There is no improvement over the matched SP cost.

10.7 COMMENSURATE DELAYS

Cost functional values for prediction schemes, where there will be two delays—the plant delay and the delay model in the controller, may be found by single-delay methods when the delays are equal, or when the delay model is replaced by a delay-free model. These are then single-delay problems. We shall demonstrate how cost functional expressions in closed form may be obtained when mismatch is such that delays are commensurate. Values of $J(h, h/2)$, and $J(h, 2h)$ will be found for the SP scheme with $CG = K/s$, assuming no mismatch in the sub-plant. These examples will serve to illlustrate the method. The corresponding stability problems were addressed in Chapter 2.

Example 10.7
In this example it is convenient to calculate $J(2H, H)$ with $E(s) = X(s) - Y(s)$, and then obtain the corresponding $J(h, h/2)$ for $e(t) = x(t) - y(t)$, for comparison with the matched value, $J(h, h)$.

$$E(s) = \frac{1}{s} \frac{s + K - Kz}{s + K + K(z^2 - z)}, \quad \text{when } x(t) = H(t), \tag{10.42}$$

$CG = K/s$, $z = \exp(-sH)$.

$$\therefore J(2H, H) = \frac{1}{2\pi i} \int_{-i\infty}^{+i\infty} -\frac{1}{s^2} \frac{(s + K - Kz)(-s + K - K\bar{z})}{(s + K + K(z^2 - z))(-s + K + K(\bar{z}^2 - Z6b))} ds \tag{10.43}$$

$$\therefore KJ(2H, H) = -\frac{1}{2\pi i} \int_{-i\infty}^{+i\infty} \frac{(p + 1 - z)(-p + 1 - \bar{z})}{p^2(p + 1 + z^2 - z)(-p + 1 + \bar{z}^2 - \bar{z})} dp, \tag{10.44}$$

using $s = Kp$, $z = \exp(-KHp)$.

Close the contour to the left, and indent at origin to the left, as the residue at the origin is zero.

The roots of $p + 1 + z^2 - z = 0$ lie within the contour by the assumption of stability using the methods of Chapter 2.

At these roots, $z^2 = z - p - 1$, which is substituted after $\bar{z}$ terms have been removed by multiplying numerator and denominator by z^2. This results in

$$KJ(2H, H) = \sum_{p: z^2 + p + 1 - z = 0} \text{res} \frac{1}{p^3} \frac{p + 1 - z}{p + 1 + z^2 - z} \frac{p^2 - 1 - pz}{(z - p)}, \tag{10.45}$$

$$= \sum_{p = 0, z - p = 0} \text{res} \frac{1}{p^3} \frac{1}{z - p} \frac{(p + 1 - z)(p^2 - 1 - pz)}{p + 1 + z^2 - z}, \tag{10.46}$$

using 'sum of *all* residues equals zero'. The poles are now at $p = 0$, and at the roots of $z - p = 0$.

The contribution from the pole at the origin is the term in p^2 in the expansion of $-(p + 1 - z)(p^2 - 1 - pz) / [(z - p)(p + 1 + z^2 - z)]$, which is $1 + 3H + 3H^2/2$. The contribution of the poles at $z - p = 0$ reduces after the substitution $z = p$ in the other terms to

$$\sum_{z - p = 0} \text{res} \frac{1}{p^3(z - p)(p^2 + 1)} = -\sum_{0, \pm i} \text{res} \frac{1}{p^3(z - p)(p^2 + 1)}, \tag{10.47}$$

i.e. the term in p^2 of $-1/[(z - p)(p^2 + 1)]$, which is $-2H - H^2/2$ together with

$$-\operatorname{Im}\left\{\frac{1}{p^3(z-p)}\right\}_{p=i} = -\frac{\cos HK}{2(1+\sin HK)}.$$

Collecting terms:

$$J(2H, H) = \frac{1}{K}\left\{1 + HK + H^2K^2 - \frac{\cos HK}{2(1+\sin HK)}\right\}. \tag{10.48}$$

Hence

$$J(h, h/2) = \frac{1}{K}\left(1 + \frac{hK}{2} + \frac{h^2K^2}{4} - \frac{\cos(hK/2)}{2(1+\sin(hK/2))}\right), \tag{10.49}$$

and using $J_{SP} = J - h$, we obtain

$$J_{SP}(h, h/2) = \frac{1}{K}\left(1 - \frac{h}{2}K + \frac{h^2K^2}{4} - \frac{\cos(hK/2)}{2(1+\sin(hK/2))}\right). \tag{10.50}$$

At $h = 0$, $J(0, 0) = 1/(2K)$, as it should.

Fig. 10.5 shows $KJ_{SP}(h, h/2)$ plotted against hK and for comparison the corresponding $KJ_{SP}(h, 0)$ given by $\cos Kh/[2(1 - \sin Kh)] - hK$. In both cases we see that for a wide range of h the mismatch in delay results in a lower cost than the value of the matched Smith predictor scheme, $J_{SP}(h, h)$.

In addition Fig. 10.5 shows the cost $J_{SP}(h, 2h)$, which is evaluated in the next example. We see that $J_{SP}(h, 2h)$ exceeds the matched cost for all positive hK.

Example 10.8
The corresponding calculation for $J(h, 2h)$, when $e(t) = x(t) - y(t)$ is given as follows:

$$E(s) = \frac{1}{s}\left(1 - \frac{Kz}{s + K + K(z - z^2)}\right)$$

and

$$J(h, 2h) = \frac{1}{2\pi i}\int_{-i\infty}^{i\infty} -\frac{1}{s^2}\left(1 - \frac{Kz}{s + K + K(z - z^2)}\right)\left(1 - \frac{K\bar{z}}{-s + K + K(\bar{z} - \bar{z}^2)}\right)ds. \tag{10.51}$$

Writing $s = Kp$, so that $z = \exp(-phK)$, $z^2 = \exp(-2phK)$ we have

$$KJ(h, 2h) = -\frac{1}{2\pi i}\int_{-i\infty}^{i\infty} \frac{1}{p^2}\left(1 - \frac{z}{1 + p + z - z^2}\right)\left(1 - \frac{\bar{z}}{1 - p + \bar{z} - \bar{z}^2}\right)dp. \tag{10.52}$$

At $p = 0$, each bracketed term is zero so that there is no singularity at $p = 0$. Close the contour to the left. Within the LH plane there are roots of $1 + p + z - z^2 = 0$, only. Hence

$$KJ(h, 2h) = -\sum_{z^2=1+p+z} \text{res} \frac{1}{p^2}\left(1 - \frac{z}{1+p+z-z^2}\right)\left(1 - \frac{z}{(1+p+z)(1-p)+z-1}\right)$$

$$= -\sum_{z^2=1+p+z} \text{res} \frac{1}{p^2}\left(1 - \frac{z}{1+p+z-z^2}\right)\left(1 - \frac{z}{-p^2+z(2-p)}\right). \quad (10.53)$$

Since the sum of *all* the residues is zero, it follows that

$$KJ(h, 2h) = \sum_{p=0, z=p^2/(2-p)} \text{res} \frac{1}{p^2}\left(1 - \frac{z}{1+p+z-z^2}\right)\left(1 - \frac{z}{-p^2+z(2-p)}\right). \quad (10.54)$$

The first bracketed term vanishes at $p = 0$, so that the residue at $p = 0$ is

$$\frac{1}{2}\frac{d}{dp}\left(\frac{-z}{1+p+z-z^2}\right)\bigg|_{p=0} = hK + \frac{1}{2}.$$

The contribution from the residues at $z = p^2/(2-p)$ is given by

$$\sum_{z=p^2/(2-p)} \text{res} \frac{1}{p^2}\left(1 - \frac{\frac{p^2}{2-p}}{1+p+\frac{p^2}{2-p}-\frac{p^4}{(2-p)^2}}\right)\left(1 - \frac{z}{-p^2+z(2-p)}\right) \quad (10.55)$$

$$= \sum_{z=p^2/(2-p)} \text{res} \frac{1}{p^2}\left(1 - \frac{p^2(2-p)}{(1+p)(2-p)^2+p^2(2-p)-p^4}\right)\left(\frac{-z}{-p^2+z(2-p)}\right) \quad (10.56)$$

$$= -\sum_{z=p^2/(2-p)} \text{res} \left\{\frac{1}{p^2} - \frac{(2-p)}{4-p^2-p^4}\right\}\left\{\frac{z}{-p^2+z(2-p)}\right\}. \quad (10.57)$$

$$= \sum_{p=0} \text{res} \frac{1}{p^2}\left(\frac{z}{-p^2+z(2-p)}\right)$$

$$+ \sum_{p=\pm\alpha, \pm i\beta} \text{res} \frac{(2-p)}{(p^2-\alpha^2)(p^2+\beta^2)}\left(\frac{z}{-p^2+z(2-p)}\right), \quad (10.58)$$

where α^2, $\beta^2 = (\sqrt{17}\pm 1)/2$ so that $\alpha^2\beta^2 = 4$, $\beta^2 - \alpha^2 = 1$ and $p^4 + p^2 - 4 = (p^2 - \alpha^2)(p^2 + \beta^2)$, and we have again the used fact that the sum of *all* the residues is zero.

Evaluating the terms for $p = 0, \pm\alpha, \pm\beta$ we obtain

$$\frac{1}{4} - \frac{1}{2\alpha(\alpha^2+\beta^2)}\left(\frac{\alpha\cosh\alpha hK + 2\sinh\alpha hK}{-\alpha^2 + 2\cosh\alpha hK + \alpha\sinh\alpha hK}\right)$$

$$+ \frac{1}{2\beta(\alpha^2+\beta^2)}\left(\frac{\beta\cos\beta hK + 2\sin\beta hK}{\beta^2 + 2\cos\beta hK - \beta\sin\beta hK}\right). \quad (10.59)$$

Collecting terms we find that

$$KJ(h, 2h) = 3/4 + \frac{\beta\cos\theta + 2\sin\theta}{2\beta(\alpha^2+\beta^2)(\beta^2 + 2\cos\theta - \beta\sin\theta)}$$

$$+ \frac{\alpha \cosh \varphi + 2\sinh \varphi}{2\alpha(\alpha^2 + \beta^2)(\alpha^2 - 2\cosh \varphi - \alpha \sinh \varphi)} + hK \qquad (10.60)$$

where $\theta = \beta hK$, $\varphi = \alpha hK$, and $\alpha^2 = (-1 + \sqrt{17})/2$, $\beta^2 = (1 + \sqrt{17})/2$, as before.

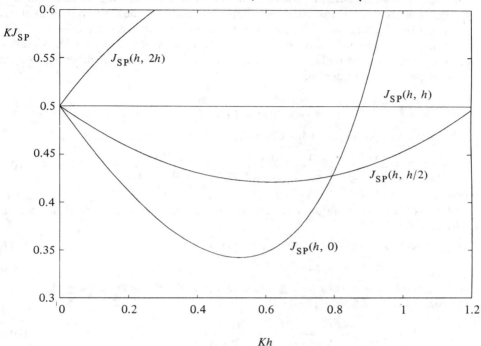

Fig. 10.5. Costs for commensurate delay problems: $CG = K/s$.

At $h = 0$, $J_{SP} = J(h, 2h) - h = 1/(2K)$, as it should, and $\partial J(h, 2h)/\partial h$ evaluated at the origin is positive, showing that for small h, at least, there is no improvement owing to this over-estimation of h_0 by a factor of 2.

The denominator term $\beta^2 + 2\cos \theta - \beta \sin \theta$ can be checked by showing that the value of θ ($=\beta hK$) at which this vanishes is consistent with the value of θ for which the real and imaginary parts of the characteristic equation $K + s + K(z - z^2)|_{s=i\beta} = 0$ also vanish.

$KJ_{SP}(h, 2h)$ is shown in Fig. 10.5, where it is seen that it always exceeds the SP delay-free value for all hK of interest.

10.8 PADÉ ALL-PASS MODEL IN A PREDICTOR SCHEME

A technique sometimes advocated for the SP scheme is to use a delay-free model for the delay in the predictor. The favoured model is a Padé approximant, usually of the all-pass type. In the example to follow, a simple all-pass model is used, the Taylor series for the exponential term is matched to second order, by the Padé all-pass model $(1 - sh_0/2)/(1 + sh_0/2) = 1 - sh_0 + s^2h_0^2/2 + O(s^3)$. It is algebraically convenient to set $h_0/2 = \alpha$ and to write $J(h, 2\alpha)$ for the corresponding cost. The resulting calculation is a single-delay problem, but the Padé approximant raises the order of the coefficients in $E(s)$, and correspondingly the order of the finite polynomial over the roots of which residues are found in the final stage of the calculation.

Denote the Padé model by $P_1 = (1 - s\alpha)/(1 + s\alpha)$, where $\alpha = h_0/2$. As we are to consider the effect of replacing the delay model by a Padé model we assume that there is parametric matching

$$E(s) = X\left(1 - \frac{CGz}{1 + CG + CGz - CGP_1}\right). \tag{10.61}$$

With $x(t)$ a step input and with $CG = K/s$ we obtain

$$E(s) = \frac{\alpha s + (1 + 2\alpha K)}{\alpha s^2 + s(1 + 2\alpha K) + K(1 + s\alpha)z} = \frac{B(s)}{A(s) + C(s)z} \tag{10.62}$$

where $A = \alpha s^2 + s(1 + 2\alpha K) = sB$ and $C = K(1 + s\alpha)$.

We shall use the techniques of section 10.6, as this is now a single-delay problem with $D = 0$.

$$\Delta(s) = A\bar{A} - C\bar{C} = \alpha^2 s^4 - s^2((1 + \alpha K)(1 + 3\alpha K)) - K^2 \tag{10.63}$$

$$= \alpha^2(s^2 - \gamma^2)(s^2 + \beta^2), \tag{10.64}$$

so that $\Delta(s)$ has two real roots $\pm\gamma$, and two purely imaginary roots $\pm i\beta$.

From section 10.6 we write

$$J = -\sum_{\Delta(s)=0} \text{res } \frac{AB\bar{B}(C - \bar{A}\bar{z})}{\Delta(s)C(A - \bar{A} + Cz - \bar{C}\bar{z})} \tag{10.65}$$

$= J_\gamma + J_\beta$, where each of these terms arises from the appropriate roots $\pm\gamma$, $\pm i\beta$.

Further we write $\Delta(s) = (s^2 - \gamma^2)\Delta_\gamma(s) = (s^2 + \beta^2)\Delta_\beta(s)$.

Using (10.35) with α replaced by γ we have

$$J_\gamma = -\frac{B\bar{B}\,\text{Ev}(A - \bar{C}\bar{z})}{2\gamma\Delta_\gamma(s)\,\text{Odd}(A + Cz)}\bigg|_{s=\gamma} \tag{10.66}$$

and

$$J_\beta = \frac{B\bar{B}\,\text{Re}(A - \bar{C}\bar{z})}{2\beta\Delta_\beta(s)\,\text{Im}(A + Cz)}\bigg|_{s=i\beta}. \tag{10.67}$$

Replacing A, B, C by their algebraic values, and exploiting obvious inter-relationships, we find the corresponding cost is given by

$$J_{\text{SP}}(h, 2\alpha) = J_\gamma + J_\beta - h$$

$$= \frac{K^2(1 - \alpha^2\gamma^2)}{2\alpha^2\gamma^3(\beta^2 + \gamma^2)} \frac{\alpha\gamma^2 - K\cosh(\gamma h) + K\alpha\gamma\sinh(\gamma h)}{\gamma(1 + 2\alpha K) + K\alpha\gamma\cosh(\gamma h) - K\sinh(\gamma h)}$$

$$+ \frac{K^2(1 + \alpha^2\beta^2)}{2\alpha^2\beta^3(\beta^2 + \gamma^2)} \frac{\alpha\beta^2 + K\cos\beta h + \alpha\beta K\sin\beta h}{\beta(1 + 2\alpha K) - K\alpha\beta\cos\beta h - K\sin\beta h} - h. \tag{10.68}$$

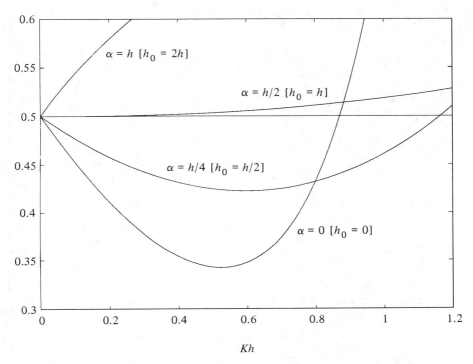

Fig. 10.6. Costs for commensurate cases with Padé predictor model $\alpha = h_0/2$.

It can be shown that, when $\alpha = 0$, the first term reduces to zero, and the second (trigonometric) term reduces to $\cos Kh/[2K(1 - \sin Kh)]$ so that $J_{SP}(h, 0) = \cos Kh/[2K(1 - \sin Kh)] - h$, as it should. Similarly, agreement with $\partial J(h, 0)/\partial h_0$ may be shown, using $\alpha = h_0/2$. It has been independently verified (Marshall, Walton and Barnes, 1991) that there is also agreement with $\partial^2 J(h, 0)/\partial h_0^2$, which is to be expected as the first Padé expansion, P_1, is itself correct to the second derivative in the expansion of $\exp(-sh_0)$.

Fig. 10.6 shows values of $J_{SP}(h, 2\alpha)$ plotted for various values of Kh, and α.

When α takes the values 0, $h/4$, $h/2$, and h in $J_{SP}(h, 2\alpha)$ we obtain the set of curves shown in Fig. 10.6. Because the cost $J_{SP}(h, 2\alpha)$ has been found for the approximant model these curves will share some, but not all, of the features of the corresponding case where the model is exactly $\exp(-sh_0)$. For the α values taken we would expect to be close (especially for low h) to the corresponding $J_{SP}(h, 0)$, $J_{SP}(h, h/2)$, $J_{SP}(h, h)$ and $J_{SP}(h, 2h)$ of Fig. 10.5.

Fig. 10.7 shows the curves of Figs. 10.6 and 10.5 together. $J_{SP}(h, h_0)_{h_0=0}$ and $J_{SP}(h, 2\alpha)_{\alpha=0}$ are indistinguishable, as they should be, as the approximant is exact (and equal to unity) at $\alpha = 0$. We see that the $J_{SP}(h, h_0)_{h_0=2h}$ and $J_{SP}(h, 2\alpha)_{\alpha=h}$, though close, are of little interest, as they exceed the SP matched cost of $\frac{1}{2}K$. The horizontal line representing $J_{SP}(h, h_0)_{h_0=h}$ lies below the $J_{SP}(h, 2\alpha)_{\alpha=h/2}$.

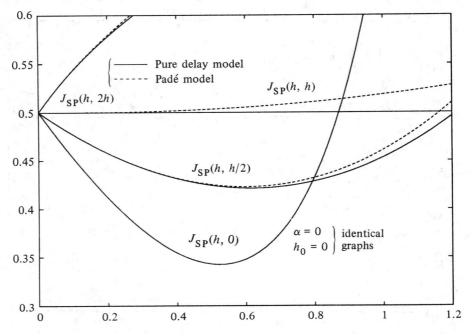

Fig. 10.7. Comparison of commensurate and Padé model costs.

The curves $J_{SP}(h, h_0)_{h_0=h/2}$ and $J_{SP}(h, 2\alpha)_{\alpha=h/4}$ are close, with very similar minima.

Fig. 10.8 shows the result of plotting the cost where, for interest only, both delays have been replaced by Padé all-pass first-order models. The expressions corresponding to the $h_0 = 0$ case, and the $h_0 = h/2$ case where every delay is modelled by the corresponding Padé expression, are given by

$$\frac{1}{2K}\left(\frac{2 - hK}{2 + hK} - 2K\right) \quad \text{and} \quad \frac{24 - 6hK + 7(hK)^2 + (hK)^3}{2K(24 + 6hK + (hK)^2)}$$

respectively. We see immediately there is poor correspondence, both for the $h_0 = 0$ case, and for the $h_0 = h/2$ case, where both the position of minima and the costs are not comparable. Poor agreement is not surprising, as these are delay-free calculations, no advantage at all having been taken, or attempted, at meeting the problems raised by exponential terms.

Fig. 10.9, for which the horizontal axis is 2α, shows some new features. Each of the curves is for the fixed value of h shown. We see that for $h = \pi/6$ the cost takes its minimum value at $\alpha = 0$, exactly equivalent to $h_0 = 0$. W see that $\partial J/\partial \alpha$ takes the value 0, at $\alpha = 0$ for $h = 0.74$, and is negative for higher values of h, so that in this Padé case $h = 0.74$ is the highest value of h for which $\alpha = 0$ gives the minimum value of cost.

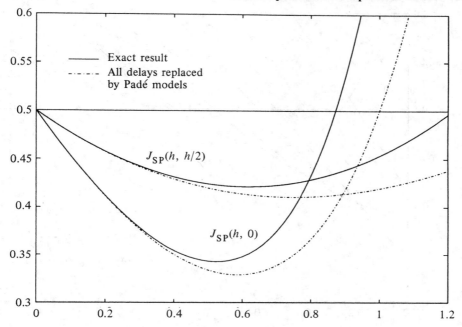

Fig. 10.8. Costs with all delays replaced by Padé models.

In the matched SP case, the cost is constant when the delay is modelled exactly (i.e. by the exponential). Where the Padé model is used we see there is a value of h of about 1.3 where the SP matched cost is always exceeded, whatever the value of α. In contrast the exact model with $h_0 = h$ would always give the SP value, which by temporal mismatch could possibly be improved upon.

Further, for low h we see the curve for a particular value of h crosses the SP constant ($=0.5$) at approximately $2\alpha = h$, as we would expect, the agreement between $\exp(-sh_0)$ and its approximant being good for low values of the argument.

This example illustrates some of the disadvantages of the use of a Padé model for the exponential. Its use is only for a restricted range of h values, beyond which the SP cost is exceeded. Use of a high-order approximant, which would need to be all-pass of course, may improve matters.

We note again that the use of totally delay-free methods for the calculation of costs is likely to be misleading, and using the approximant as an approximate way of finding costs has no guarantee of success. It is surely best to find the exact result where this is possible. The method given in section 10.6 makes this possible for a wide range of problems, as the problem has been reduced to finding odd and even parts of functions evaluated over the roots of a polynomial. Where this polynomial is such that the roots need to be found numerically, the method still gives formulae in which h, the delay value, is explicit, i.e. one numerical calculation of roots for all values of h.

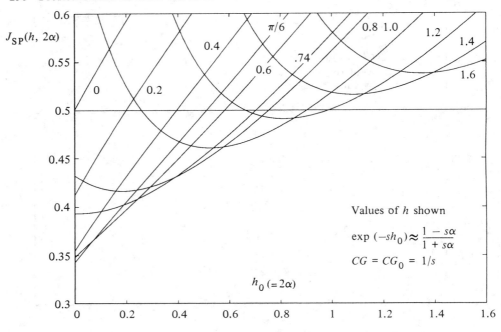

Fig. 10.9. Costs vs delay $h_0 = 2\alpha$ for various h.

10.9 SUMMARY

The examples in this chapter show how the methods introduced earlier, and refined here where appropriate, may be applied to stability and evaluation of performance of a particular class of prediction control scheme. Particular temporal and parametric mismatch problems have been given. It is possible to extend these methods to plants with CG of higher order than we have given here, and to systems with delays related by integers. There are some new results here, and some surprising insights into the applicability or otherwise of the Smith predictor method of plants of low order, and with small delays.

The methods given in this chapter, though related to SP control, now better understood, are of course more widely applicable. In particular it is hoped that the new methods met here may help in on-line control of time-delay systems, given that it is now possible to compute cost functions quickly from the algebraic expression, rather than finding points in simulation followed by interpolation.

It is expected that the closed-form expression given here, and the solutions that may be found by others using these methods, may be found useful in other areas of control, and theory.

It has been the case hitherto that analytical solutions were not available for many time-delay problems owing to the need for summing an infinite number of residues. These problems have now been reduced to summing a finite number of residues. In some cases, where the cost functional is a sum of a finite number of terms, it may be possible to reduce the number of terms further. For example, where there are hyperbolic and trigonometric terms it is sometimes the case that one or other (not necessarily the trig. terms) dominate the expression. So we see that the problem of reduction to a finite number of terms may, on occasion, be reduced even further.

It is expected that cost functionals (and the associated stability problems) will be extended to systems of higher order, where the algebraic burden increases somewhat, by the use of computer algebra, and some steps have been taken in this direction (Marshall, Walton and Barnes, 1992).

It is sometimes suggested that off-diagonal Padé models might be used to model the delay h_0 in prediction schemes, when there is noise present. It is now possible to treat this analytically by the methods given here, and also of course to extend the example given in section 10.8 to diagonal all-pass models of second and higher order. We note that the cost functional for the second-order case would agree in its sensitivity coefficients w.r.t. h_0 up to 4th order, at the expense of the increased algebraic calculation. We note of course that the algebraic problems are 'one-off'; once the calculation is done (and, as we have stressed, verified) one has an algebraic expression upon which the methods of analysis and approximation may then be applied if appropriate. Again, computer algebra, properly understood, could be applied to such examples.

The interpretation of the various terms of the cost functional in terms of energy, and the location of the roots of $\Delta(s) = A\bar{A} - C\bar{C}$, including as they do the roots of interest in the stability problem, is another interesting open problem. One may ask: what is the significance of the *real* roots which appear to be associated with complex values of delay. These terms are important in finding the cost functionals although they do not contribute directly to the stability problem, in any obvious way.

The Smith predictor methods are closely related (see, for example, Morari and Zafiriou (1989)) to 'internal model control'. It might be expected that many of the present methods might find application there also. These latter applications chapters are related to the work of the authors and it would be surprising if the new methods described herein were not immediately applicable for time-delay control schemes and methods which rival those given here.

References

Allwright, D.J. (1980) A note on Routh–Hurwitz determinants and integral square errors, *Int. J. Control* **31**(4), 807–810.

Åström, K.J. (1968) Recursive formulas for the evaluation of certain complex integrals, Rep. 6804, Lund Institute of Technology, Division of Automatic Control.

Åström, K.J. (1970) *Introduction to Stochastic Control Theory*, Academic Press.

Åström, K.J. and Wittenmark, B. (1984) *Computer Controlled Systems: Theory and Design,* Prentice-Hall International.

Baker, G.A. and Graves-Morris, P.R. (1981) *Padé Approximants*, Vol. 1, Addison-Wesley.

Barker, R.H. (1952) The pulse-transfer function and its application to sampling servo-systems, *Proc. IEE* **99**, Pt.4. 302–317.

Barnett, S. (1975) *Introduction to Mathematical Control Theory*, Clarendon Press, Oxford.

Barnett, S. (1984) *Matrices in Control Theory* (Revised Edition), Robert Krieger Publishing Co., Malabar, Florida.

Bateman, H. (1945) The control of an elastic fluid, *Brit. Am. Math. Soc.* **51**, 601–646.

Bellman, R. and Kalaba, R. (1964) *Mathematical Trends in Control Theory*, Dover.

Burnside, W.S. and Panton, A.W. (1960) *The Theory of Equations*, Vol. II, Dover Publications.

Byron, R., Cox, C.S. and Ball, D.J. (1979) An application of a Smith controller in a Sinter plant, *IEE Colloquium Digest*, 1979/38.

Castelan, W.B. and Infante, E.F. (1977) On a functional equation arising in the stability theory of difference–differential equation, *Quarterly of Appl. Math.* **35**(3), 311–319.

Castelan, W.B. and Infante, E.F. (1979) A Liapunov functional for a matrix neutral difference–differential equation with one delay, *J. Math. Anal. Appl.* **71**(1), 105–130.

Choksy, N.H. (1960) Time-lag systems—a bibliography, *IEEE Trans. on Automatic Control* AC-5, 66–70.
Chotai, A. (1980) Identification errors and the control of time-delay systems, Ph.D. Thesis, University of Bath.
Churchill, R.V., Brown, J.W. and Verhay, R.F. (1974) *Complex Variables and Applications,* 3rd Edn, McGraw-Hill Kogakusha.
Datko, R. (1972) An algorithm for computing Liapunov functionals for some differential difference equations, in *Ordinary Differential Equations,* 1971 NRL-MRC Conference. Academic Press, 387–398.
Delfour, M.C., McCalla, C. and Mitter, S.K. (1975) Stability and the infinite-time quadratic cost problem for linear hereditary differential systems, *SIAM J. Control* **13**(1), 48–88.
Fifer, S. (1961) *Analogue Computation,* Vol. IV, 1239–1293, McGraw-Hill.
Fuller, A.T. (1967) The replacement of saturation constraints by energy constraints in control optimization theory, *Int. J. Control* **6**(3), 201–207.
Gantmacher, F.R. (1964) *The Theory of Matrices,* Vol. 2, Chelsea Publishing Co., N.Y.
Garland, B. and Marshall, J.E. (1974) Sensitivity considerations of Smith's method for time-delay systems, *Electr. Lett.* **10**(15), 308–309.
Gawthrop, P.J. (1977) Some interpretations of the self-tuning controller, *Proc. IEE.* **124**(10), 88–894.
Gorecki, H., Fuksa, S., Grabowski, P. and Korytowski, A. (1989) *Analysis and Synthesis of Time-delay Systems,* John Wiley and Sons, and PWN—Warsaw.
Gorecki, H., Fuksa, S., Korytowski, A. and Mitkowski, W. (1983) *Sterowanie optymalne w systemach liniowych z kwadratowym wskaznikiem jakoski,* Panstwowe Wydawnictwo Naukowe, Warszawa.
Gorecki, H. and Popek, L. (1983) Control of systems with time-delay, *Third IFAC Symposium on Control of Distributed Parameter Systems,* Eds Barbary, J-P. and LeLetty, L., Pergamon.
Gorecki, H. and Popek, L. (1984) Parametric optimisation problem for control systems with time-delay, *Proc. 9th IFAC World Congress,* Budapest.
Grabowski, P. (1983) A Lyapunov functional approach to a parametric optimization for a class of infinite-dimensional control systems, *Elektrotechnika* **2** (No. 3), 207–232.
Grabowski, P. (1989) Evaluation of quadratic cost functionals for neutral systems: the frequency domain approach, *Int. J. Control* **49**(3), 1033–1053.
Graham, D. and Lathrop, R.C. (1953) The synthesis of optimum transient response, *Trans. AIEE Pt. II Appns Ind.* **72**, 273–288.
Hertz, D., Jury, E.I. and Zeheb, E. (1984) Simplified analytical stability test for systems with commensurate time-delays, *Proc. IEE* **131**, 52–56.
Hocken, R.D. and Marshall, J.E. (1982) Mismatch and the optimal control of linear systems with time-delays, *Opt. Control Applns and Methods* **3**, 211–219.
Hocken, R.D. and Marshall, J.E. (1983) The effects of mismatch on an optimal control scheme for linear system with control time-delays, *Opt. Control Applns and Methods* **4**, 47–69.

Hocken, R.D., Marshall, J.E. and Salehi, S.V. (1983) Time-delay control: mismatch problems, *Third IFAC Symposium on Control of Distributed Parameter Systems*, eds Barbary, J-P. and LeLetty, L., Pergamon.

Hocken, R.D., Salehi, S.V. and Marshall, J.E. (1983) Time-delay mismatch and the performance of prediction control schemes, *Int. J. Control* **38**(2), 433–447.

Hydon, P.E., Marshall, J.E. and Walton, K. (1992) All-pass feedback systems, *6th Intl Conference on Control Theory, Glasgow*, O.U.P.

Infante, E.F. and Castelan W.B. (1978) A Liapunov functional for a matrix difference-differential equation, *J. Differential Equations* **29**, 439–451.

Jacobs, O.L.R. (1974) *Introduction to Control Theory*, Clarendon Press, Oxford.

James, H.M., Nichols, N.B. and Phillips, R.S. (1947) *Theory of Servo Mechanisms*, M.I.T. Radiation Lab. Series, Vol. 25, McGraw-Hill.

Jury, E.I. (1958) *Sampled-data Control Systems*, John Wiley.

Jury, E.I. (1964) On the roots of real polynomial inside the unit circle and a stability criterion for linear discrete systems, *Proc. of Sec. Cong. of IFAC*, Butterworths, London.

Jury, E.I. (1965) A note on the evaluation of the total square integral, *IEEE Trans. on Automatic Control (Correspondence) AC-10*, 110–111.

Jury, E.I. (1974) *Inners and the Stability of Dynamic Systems*, John Wiley.

Kalman, R.E. and Bertram, J.E. (1960) Control system analysis and design via the second method of Lyapunov, *ASMEJ. Basic Eng.* **82**, 371–400.

Korn, G.A.T. and Korn, T.M. (1968) *Mathematical Handbook for Scientists and Engineers*, McGraw-Hill.

Krall, A.M. (1967) *Stability Techniques for Continuous Linear Systems*, Gordon and Breach.

Kuo, B. (1980) *Digital Control Systems*, Holt-Saunders.

Kwakernaak, H. and Sivan, R. (1972) *Linear Optimal Control Systems*, Wiley-Interscience, N.Y.

Levin, B. (1964) *Distribution of zeros of entire functions*, A.M.S. Trans. of Math., Monograph No. 5.

MacDonald, N. (1989) *Biological Delay Systems: Linear Stability Theory*, Cambridge University Press.

MacFarlane, A.G.J. (1963) The calculation of functionals of the time and frequency response of a linear constant coefficient dynamical system, *Quart. J. Mech. Appl. Math.* **16**, 259–271.

MacFarlane, A.G.J. (1970) Multivariable control system design: A guided tour, *Proc. IEE* **117**(5), 1039–47.

Marshall, J.E. (1974) Extensions of O.J. Smith's method to digital and other systems, *Int. J. Control* **19**(5), 933–939.

Marshall, J.E. (1979) *Control of Time Delay Systems*, Peter Peregrinus, IEE, London.

Marshall, J.E. and Salehi, S.V. (1982) Improvement of performance by the use of time-delay elements, *Proc. IEE*. (Pt. D) **129**(5), 177–181.

Marshall, J.E., Walton, K. and Barnes, A. (1991) Cost functionals for non-commensurate delay systems, *6th International IMA conf. on Control Theory, Glasgow*, O.U.P.

Morari, M. and Zafiriou, E. (1989) *Robust Process Control*, Prentice-Hall.
Naslin, P. (1968) *Essentials of Optimal Control*, Iliffe Books, London.
Neimark, J.I. (1949) *Stability of Linear Systems*, Leningrad.
Newton, G.C., Gould, L.A. and Kaiser, H. (1957) *Analytical Design of Linear Feedback Controls*, John Wiley.
Ogata, K. (1987) *State Space Analysis of Control Systems*, Prentice-Hall.
Padé, H. (1892) Sur la representation approchée d'une fonction par les fractions rationelles, *Ann. de L'école Normale Sup.*, 3ieme Serie **9**, Suppl. 3–93.
Padé, H. (1899) Memoire sur les developements en fractiones continues de la fonction exponentielle etc., *Ann. Sci. École. Norm. Sup.* **16**, 395–426.
Pontryagin, L. (1955) On the zeros of some elementary transcendental functions, *Amer. Math. Soc. Trans.* **2**, 95–110.
Popov, E.P. (1962) *The Dynamics of Automatic Control Systems*, Pergamon.
Ralston, A. and Wilf, H.S. (1965) *Mathematical Methods for Digital Computers*, John Wiley.
Repin, Yu. M. (1965) Quadratic Liapunov functionals for systems with delay, *Prikladnaya Matematika i Mekhanika* **29**, No. 3, 564–566. Translated as *J. Appl. Math. Mech.* **29**, 669–672.
Rosenbrock, H.H. (1970) *State-space and Multivariable Theory*, Nelson.
Rosenbrock, H.H. (1974) *Computer Aided Control System Design*, Academic Press.
Rosenbrock, H.H. and Storey, C. (1970) *Mathematics of Dynamical Systems*, Nelson.
Sklansky, J. (1958) On closed form expressions for mean squares in discrete-continuous systems, *IRE Trans. on Automatic Control* **AC-3**, 21–27.
Smith, O.J.M. (1957) Closer control of loops with dead time, *Chem. Eng. Prog. Transactions* **53**(5), 216–219.
Smith, O.J.M. (1959) A controller to overcome dead time, *I.S.A.J.* **6**(2), 28–33.
Stépán, G. (1989) *Retarded Dynamical Systems: Stability and Characteristic Funtions*, Pitman Research Notes in Mathematics Series, Longman Scientific and Technical.
Storer, J.E. (1957) *Passive Network Synthesis*, McGraw-Hill.
Thowsen, A. (1981) An analytical stability test for a class of time-delay systems, *IEEE Trans. on Automatic Control* **AC-26**, 735–736.
Truxal, J.G. (1958) *Automatic Feedback Control System Synthesis*, McGraw-Hill.
Wall, H.S. (1948) *Analytical Theory of Continued Fractions*, Van Nostrand.
Walton, K. and Gorecki, H. (1984) On the evaluation of cost functionals with particular emphasis on time-delay systems, *IMA J. Math. Control and Infn.* **I**(3), 283–306.
Walton, K. and Marshall, J.E. (1984a) Closed-form solution for time-delay systems cost functionals, *Int. J. Control* **39**(5), 1063–1071.
Walton, K. and Marshall, J.E. (1984b) Mismatch in a predictor control scheme: some closed-form solutions, *Int. J. Control* **40**, 403–419.
Walton, K., Ireland, B. and Marshall, J.E. (1986) Evaluation of weighted quadratic functionals for time-delay systems, *Int. J. Control* **44**(6), 1491–1498.
Walton, K. and Marshall, J.E. (1987a) Direct method for TDS stability analysis, *Proc. IEE* **134** pt(D), 101–107.
Walton, K. and Marshall, J.E. (1987b) Evaluation of cost functionals for systems with commensurate delays by a finite polynomial method, *Proc. IEE* **134** pt(D), 108–114.

Index

accessibility for control, 202
addition of delay, deliberate, 210
advanced system, 19, 21
all-pass functions, 143
  property, 144
  systems, 143
    cost-difference theorem, 152–155
    cost function evaluation, 148–152
    exponential inputs, 155–156
    stability, 145–148
    sgn relationship for, 145
    slope rule for, 147
analytical design, 2, 213
arbitrary initial conditions, 75, 101–104
arbitrary inputs, 75, 77–82
asymptotic stability, condition for, 20

Cauchy problem, 101
characteristic equation, 19, 93, 204
commensurate delays, 17, 30–37, 110–112, 227–231
  two, 59–64, 227
  several, 65–71
common roots of $A(s), C(s)$, 56
contour at infinity, 56
conventional system, 162
  driven by white noise, 196–200
convolution theorem of the Laplace transform, 4
cost criterion, 2
cost-difference theorem, 152–155
cost functionals, *see* performance criteria

counterflow equations, 87, 91
covariance matrix, 192
criteria, weighted quadratic, 48, 71, 104
  by exponential, 50, 105
  by powers of t, 48, 73, 106
cross-sensitivity, 211
crossing points, 21
  stabilizing, 34
  destabilizing, 34
deformed contour, 55
delay theorem, of the Laplace transform, 18
  of z-transform, 124
delay-free systems, 3–14
destabilizing values of ω, 23–24
destabilizing zeros, 21
digital simulation, 125
direct method for TDS stability, 14, 19–28
discrete controller, 121

error, 2
error, Laplace transform of, 8
error, transform, general form of, 19, 55
  for two commensurate delays, 62
  for several commensurate delays, 65
evaluation of J when $D(s)=0$, 224–227
exponential function, and Padé approximant, 156–157
exponential inputs to all-pass systems, 155–156

feedback cost functional, 144
fictitious delay, 124

forward-path cost functional, 144
fractional delay, 124
function stored in the delay, 17, 75–77, 81–82
  as an input, 77
fundamental matrix, 5, 88, 195

gain function, 143
Gaussian white noise, 173

Heaviside expansion formula, 4, 38
  generalized, 38
hold, zero-order, 122
homogeneous to non-homogeneous
  transformation, 117
hybrid systems, 121

improvement of delay-free system, 207
impulse response, 3, 5
initial conditions, 75–76, 101–104
  activation by, 180
  arbitrary, 101–104
  constant, 103
  dependence on, 184
  general, 101
  pointwise, 89
integral criteria, 2
  of squared error (ISE), 2
integrals of mixed type, evaluation of,
  135–142
integrand, separation of, 53, 57
inter-sample behaviour, 126
interface theorem, 123
internal model, 213
ISE (integral of squared error), 2
Ito calculus, 192

Jordan canonical form, 94–95

Laplace transform of error, 8
Lyapunov equation, 12
  functionals, 87
  method, 11–14, 83–88
Lyapunov systems of equations, 87
  solution of, 91

matrix description of the control system, 4,
  83
method of steps, 17–18
  for commensurate delays, 37
mismatch, 202, 204
  and performance, 205–207
  improvement by, 206, 223
  parametric, 202, 210–211
  problems, 202

  sign of, 207–210
  temporal, 202, 207–210
mixed derivative, 211
model delay, 202
modified z-transform, 124, 139
multiple commensurate delays, 32

neutral system(s), 19, 24, 88, 112–113, 165

optimal controller settings, 175
  P-controller, 175, 183
  PD-controller, 175, 183
  PI-controller, 177
  PID-controller, 178
optimization of conventional control
  systems, 172–189
output integral, 139

Padé approximant, 29, 144
  and cost functional, 160
  of an exponential, 156
  and stability, 157
Padé table, 156
  diagonal entries of, 156
parametric mismatch, 202, 204, 210–211
  cost, 217–218
  stability, 215–217
parametric sensitivity coefficient, 210, 212
Parseval's theorem
  discrete version of, 129
  for the Laplace transform, 8–9
  for the z-transform, 125
Parseval's theorem and contour integration,
  51–52
partial fractions, 41
performance and stability, 3
performance criteria, 2
  evaluation
  delay-free systems
    complex variable methods, 8–11
    matrix methods, 11–14
    time-domain methods, 6–8
  single time-delay systems
    contour integration, 52–59
    generalized Heaviside expansion, 39–47
    the Lyapunov method, 89–101
  commensurate delay systems, 59–71,
    110–112
  with weighting functions and derivatives,
    48, 71, 104
  for sampled-data systems, 125, 135
  for all-pass systems, 148–152
PI-controller, 203
PI-regulator, 86

PID-control, 162
 parameters, 163
polynomials $A(s)$ to $D(s)$, degrees of 43, 45
practical stability, 28
prediction conditions for, 202
predictive control, 201, 214
predictor scheme with Padé model, 231–234
pure time delay, 16

reduction of J_{SP} by mismatch, 206
regions of instability, 21
residues, theory of, 40, 54
retarded system, 19, 21, 165
Routh–Hurwitz criteria, 6

sampled-data systems, 121
 cost functionals for, 125–126
 stability of, 126–129
Schur–Cohn criteria, 127
sensitivity and sign of mismatch, 207
sgn relationship
 for systems with a single time delay, 23–24
 for systems with several commensurate delays, 34–36
 for all-pass systems, 145
simplification of an expression for J, 61
single-delay problem, simple example, 39–42, 52–54
 general case, 42–47, 55–59
single-delay stability, 19–26
Sklansky technique, 130
 extended, 131–135
slope rule, for systems with a single time delay, 24
 for systems with several commensurate delays, 36
 for all-pass systems, 147
Smith predictor, 202
 applicability, 223
 cost, 205
Smith's principle, 205
stability, 3
 for delay-free systems, 6
 for systems with a single time delay, 19–30
 for neutral systems, 24–25
 for systems with several commensurate delays, 30–37

independent of delay, 26, 215, 216
 windows, 27, 29, 147, 158
 for sampled-data systems, 126–129
 for all-pass systems, 145–148
stabilizing crossing points, 34
 effect of delay, 27
 values of ω, 23
 zeros, 21
stochastic inputs, 190–196
 differential equations, 192
 integral, 191
 processes, 190
 systems, 190–191
stored function, 75–77, 81–82
sub-plant model, 202
system poles, 20
system, retarded, 19
 advanced, 19
 neutral, 19
systems with many delays, 110

temporal mismatch, 202, 204, 207–210
 $h_0 = 0$, 221–224
 extreme, 214
 stability, 218–221
temporal sensitivity coefficient, 208
 table of, 212
time-delay, 16
transfer function, 4
 of pure delay, 18
transmission line problems, 30

values of ω, stabilizing, 23–24
 destabilizing, 23–24

W-polynomial, 22–27
Weierstrass decomposition, 41
weighted quadratic criteria, 48, 71, 104
weighting function, 104
white Gaussian noise, 173, 191
Wiener process, 192–196

z-transfer function, 123
z-transform, 122, 123
 delay theorem for, 124
 interface theorem, 123
 of output sequence, 123
 Parseval theorem for, 125

MATHEMATICS AND ITS APPLICATIONS
Series Editor: G. M. BELL
Emeritus Professor of Mathematics, King's College London, University of London

Farrashkhalvat, M. & Miles, J.P.	TENSOR METHODS FOR ENGINEERS AND SCIENTISTS
Faux, I.D. & Pratt, M.J.	COMPUTATIONAL GEOMETRY FOR DESIGN AND MANUFACTURE
Firby, P.A. & Gardiner, C.F.	SURFACE TOPOLOGY: Second Edition
Gardiner, C.F.	MODERN ALGEBRA
Gardiner, C.F.	ALGEBRAIC STRUCTURES
Gasson, P.C.	GEOMETRY OF SPATIAL FORMS
Gilbert, R.P. & Howard, H.C.	ORDINARY AND PARTIAL DIFFERENTIAL EQUATIONS WITH APPLICATIONS
Goodbody, A.M.	CARTESIAN TENSORS
Graham, A.	KRONECKER PRODUCTS AND MATRIX CALCULUS: with Applications
Graham, A.	MATRIX THEORY AND APPLICATIONS FOR ENGINEERS AND MATHEMATICIANS
Graham, A.	NONNEGATIVE MATRICES AND APPLICABLE TOPICS IN LINEAR ALGEBRA
Griffel, D.H.	APPLIED FUNCTIONAL ANALYSIS
Griffel, D.H.	LINEAR ALGEBRA AND ITS APPLICATIONS: Vol. 1, A First Course; Vol. 2, More Advanced
Guest, P.B.	LAPLACE TRANSFORMS AND AN INTRODUCTION TO DISTRIBUTIONS
Hanyga, A.	MATHEMATICAL THEORY OF NON-LINEAR ELASTICITY
Hart, D. & Croft, A.	MODELLING WITH PROJECTILES
Hoskins, R.F.	GENERALISED FUNCTIONS
Hoskins, R.F.	STANDARD AND NONSTANDARD ANALYSIS
Hoskins, R.F. & Sousa Pinto, J.J.M.	DISTRIBUTIONS, ULTRADISTRIBUTIONS AND OTHER GENERALISED FUNCTIONS
Hunter, S.C.	MECHANICS OF CONTINUOUS MEDIA, 2nd (Revised) Edition
Huntley, I. & Johnson, R.M.	LINEAR AND NONLINEAR DIFFERENTIAL EQUATIONS
Irons, B.M. & Shrive, N.G.	NUMERICAL METHODS IN ENGINEERING AND APPLIED SCIENCE
Ivanov, L.L.	ALGEBRAIC RECURSION THEORY
Johnson, R.M.	THEORY AND APPLICATIONS OF LINEAR DIFFERENTIAL AND DIFFERENCE EQUATIONS
Johnson, R.M.	CALCULUS: Theory and Applications in Technology and the Physical and Life Sciences
Jones, R.H. & Steele, N.C.	MATHEMATICS IN COMMUNICATION THEORY
Jordan, D.	GEOMETRIC TOPOLOGY
Kelly, J.C.	ABSTRACT ALGEBRA
Kim, K.H. & Roush, F.W.	APPLIED ABSTRACT ALGEBRA
Kim, K.H. & Roush, F.W.	TEAM THEORY
Kosinski, W.	FIELD SINGULARITIES AND WAVE ANALYSIS IN CONTINUUM MECHANICS
Krishnamurthy, V.	COMBINATORICS: Theory and Applications
Livesley, K.	MATHEMATICAL METHODS FOR ENGINEERS
Lord, E.A. & Wilson, C.B.	THE MATHEMATICAL DESCRIPTION OF SHAPE AND FORM
Malik, M., Riznichenko, G.Y. & Rubin, A.B.	BIOLOGICAL ELECTRON TRANSPORT PROCESSES AND THEIR COMPUTER SIMULATION
Marshall, J.E., Gorecki, H., Korytowski, A. & Walton, K.	TIME-DELAY SYSTEMS: Stability and Performance Criteria with Applications
Martin, D.	MANIFOLD THEORY: An Introduction for Mathematical Physicists
Massey, B.S.	MEASURES IN SCIENCE AND ENGINEERING
Meek, B.L. & Fairthorne, S.	USING COMPUTERS
Menell, A. & Bazin, M.	MATHEMATICS FOR THE BIOSCIENCES
Mikolas, M.	REAL FUNCTIONS AND ORTHOGONAL SERIES
Moore, R.	COMPUTATIONAL FUNCTIONAL ANALYSIS
Murphy, J.A., Ridout, D. & McShane, B.	NUMERICAL ANALYSIS, ALGORITHMS AND COMPUTATION
Niss, M., Blum, W. & Huntley, I.	TEACHING OF MATHEMATICAL MODELLING AND APPLICATIONS
Nonweiler, T.R.F.	COMPUTATIONAL MATHEMATICS: An Introduction to Numerical Approximation
Ogden, R.W.	NON-LINEAR ELASTIC DEFORMATIONS
Oldknow, A.	MICROCOMPUTERS IN GEOMETRY
Oldknow, A. & Smith, D.	LEARNING MATHEMATICS WITH MICROS
O'Neill, M.E. & Chorlton, F.	IDEAL AND INCOMPRESSIBLE FLUID DYNAMICS
O'Neill, M.E. & Chorlton, F.	VISCOUS AND COMPRESSIBLE FLUID DYNAMICS
Page, S.G.	MATHEMATICS: A Second Start
Prior, D. & Moscardini, A.O.	MODEL FORMULATION ANALYSIS
Rankin, R.A.	MODULAR FORMS
Scorer, R.S.	ENVIRONMENTAL AERODYNAMICS
Shivamoggi, B.K.	STABILITY OF PARALLEL GAS FLOWS
Smitalova, K. & Sujan, S.	DYAMICAL MODELS IN BIOLOGICAL SCIENCES
Srivastava, H.M. & Owa, S.	UNIVALENT FUNCTIONS, FRACTIONAL CALCULUS, AND THEIR APPLICATIONS
Stirling, D.S.G.	MATHEMATICAL ANALYSIS
Sweet, M.V.	ALGEBRA, GEOMETRY AND TRIGONOMETRY IN SCIENCE, ENGINEERING AND MATHEMATICS
Temperley, H.N.V.	GRAPH THEORY AND APPLICATIONS
Temperley, H.N.V.	LIQUIDS AND THEIR PROPERTIES
Thom, R.	MATHEMATICAL MODELS OF MORPHOGENESIS

MATHEMATICS AND ITS APPLICATIONS
Series Editor: G. M. BELL,
Emeritus Professor of Mathematics, King's College London, University of London

Toth, G.	HARMONIC AND MINIMAL MAPS AND APPLICATIONS IN GEOMETRY AND PHYSICS
Townend, M.S.	MATHEMATICS IN SPORT
Townend, M.S. & Pountney, D.C.	COMPUTER-AIDED ENGINEERING MATHEMATICS
Twizell, E.H.	COMPUTATIONAL METHODS FOR PARTIAL DIFFERENTIAL EQUATIONS
Twizell, E.H.	NUMERICAL METHODS, WITH APPLICATIONS IN THE BIOMEDICAL SCIENCES
Vein, R. & Dale, P.	DETERMINANTS AND THEIR APPLICATIONS IN MATHEMATICAL PHYSICS
Vince, A. and Morris, C.	DISCRETE MATHEMATICS FOR COMPUTING
Walton, K., Marshall, J., Gorecki, H. & Korytowski, A.	CONTROL THEORY FOR TIME DELAY SYSTEMS
Warren, M.D.	FLOW MODELLING IN INDUSTRIAL PROCESSES
Webb, J.R.L.	FUNCTIONS OF SEVERAL REAL VARIABLES
Willmore, T.J.	TOTAL CURVATURE IN RIEMANNIAN GEOMETRY
Willmore, T.J. & Hitchin, N.	GLOBAL RIEMANNIAN GEOMETRY

Statistics, Operational Research and Computational Mathematics
Editor: B. W. CONOLLY,
Emeritus Professor of Mathematics (Operational Research), Queen Mary College, University of London

Abaffy, J. & Spedicato, E.	ABS PROJECTION ALGORITHMS: Mathematical Techniques for Linear and Nonlinear Equations
Beaumont, G.P.	INTRODUCTORY APPLIED PROBABILITY
Beaumont, G.P.	PROBABILITY AND RANDOM VARIABLES
Beaumont, G.P. & Knowles, J.D.	STATISTICAL TESTS: An Introduction with Minitab Commentary
Bunday, B.D.	STATISTICAL METHODS IN RELIABILITY THEORY AND PRACTICE
Conolly, B.W.	TECHNIQUES IN OPERATIONAL RESEARCH: Vol. 1, Queueing Systems
Conolly, B.W.	TECHNIQUES IN OPERATIONAL RESEARCH: Vol. 2, Models, Search, Randomization
Conolly, B.W.	LECTURE NOTES IN QUEUEING SYSTEMS
Conolly, B.W. & Pierce, J.G.	INFORMATION MECHANICS: Transformation of Information in Management, Command, Control and Communication
Doucet, P.G. & Sloep, P.B.	MATHEMATICAL MODELING IN THE LIFE SCIENCES
French, S.	SEQUENCING AND SCHEDULING: Mathematics of the Job Shop
French, S.	DECISION THEORY: An Introduction to the Mathematics of Rationality
Goult, R.J.	APPLIED LINEAR ALGEBRA
Griffiths, P. & Hill, I.D.	APPLIED STATISTICS ALGORITHMS
Griffiths, H.B. & Oldknow, A.	MATHEMATICAL MODELS OF CONTINUOUS AND DISCRETE DYNAMIC SYSTEMS
Hartley, R.	LINEAR AND NON-LINEAR PROGRAMMING
Janacek, G. & Swift, L.	TIME SERIES: Forecasting, Simulation, Applications
Jolliffe, F.R.	SURVEY DESIGN AND ANALYSIS
Jolliffe, I.T. & Jones, B.	STATISTICAL INFERENCE
Kapadia, R. & Andersson, G.	STATISTICS EXPLAINED: Basic Concepts and Methods
Lindfield, G. & Penny, J.E.T.	MICROCOMPUTERS IN NUMERICAL ANALYSIS
Lootsma, F.	OPERATIONAL RESEARCH IN LONG TERM PLANNING
Moscardini, A.O. & Robson, E.H.	MATHEMATICAL MODELLING FOR INFORMATION TECHNOLOGY
Moshier, S.L.B.	METHODS AND PROGRAMS FOR MATHEMATICAL FUNCTIONS
Norcliffe, A. & Slater, G.	MATHEMATICS OF SOFTWARE CONSTRUCTION
Oliveira-Pinto, F.	SIMULATION CONCEPTS IN MATHEMATICAL MODELLING
Ratschek, J. & Rokne, J.	NEW COMPUTER METHODS FOR GLOBAL OPTIMIZATION
Schendel, U.	INTRODUCTION TO NUMERICAL METHODS FOR PARALLEL COMPUTERS
Schendel, U.	SPARSE MATRICES
Sehmi, N.S.	LARGE ORDER STRUCTURAL EIGENANALYSIS TECHNIQUES: Algorithms for Finite Element Systems
Sewell, G.	COMPUTATIONAL METHODS OF LINEAR ALGEBRA
Sharma, O.P.	MARKOVIAN QUEUES
Smith, D.K.	DYNAMIC PROGRAMMING: A Practical Introduction
Späth, H.	MATHEMATICAL SOFTWARE FOR LINEAR REGRESSION
Stoodley, K.D.C.	APPLIED AND COMPUTATIONAL STATISTICS: A First Course
Stoodley, K.D.C., Lewis, T. & Stainton, C.L.S.	APPLIED STATISTICAL TECHNIQUES
Thomas, L.C.	GAMES, THEORY AND APPLICATIONS
Vajda, S.	FIBONACCI AND LUCAS NUMBERS, AND THE GOLDEN SECTION
Whitehead, J.R.	THE DESIGN AND ANALYSIS OF SEQUENTIAL CLINICAL TRIALS: Second Edition
Woodford, C.	SOLVING LINEAR AND NON-LINEAR EQUATIONS

UNIVERSITY OF STRATHCLYDE

30125 00440059 3

ML

Books are to be returned on or before
the last date below.

- 9 JUN 1998
2 6 AUG 1998
1 2 JAN 2010
2 1 NOV 2001
DUE
2 1 MAR 2006
-1 OCT 2002
DUE
0 7 JAN 2009
DUE
1 7 FEB 2009

LIBREX—